Disinformation in Central Europe

Michał Wenzel • Karina Stasiuk-Krajewska
Editors

Disinformation in Central Europe

Actors, structure, impact

palgrave
macmillan

Editors
Michał Wenzel
Dept. of Social Sciences
SWPS University
Warsaw, Poland

Karina Stasiuk-Krajewska
Faculty od Law and Communication
SWPS University
Wrocław, Poland

ISBN 978-3-031-76870-5 ISBN 978-3-031-76871-2 (eBook)
https://doi.org/10.1007/978-3-031-76871-2

This Palgrave Macmillan imprint is published by the registered company Springer Nature Switzerland AG.
The registered company address is: Gewerbestrasse 11, 6330 Cham, Switzerland

Acknowledgment

The research and analysis presented in the book was carried out as part of the Central European Digital Media Observatory (CEDMO). CEDMO is an international consortium consisting of research institutions, educational institutions and media entities from Central Europe (from the Czech Republic, Poland and Slovakia), which researches and counteracts disinformation in its broadest sense. The leader of the consortium is Charles University from Prague.

Three professional groups work together within CEDMO: educators, fact-checkers and researchers. Consequently, CEDMO's activities are carried out on three levels: education, fact-checking, and research by scientists mainly in the social sciences.

This research aims to gain an in-depth and multifaceted understanding of disinformation phenomena using scientific methods. The research uses diverse methodologies and addresses different areas: senders and recipients of disinformation messages and structures of this type of content. The analyses are interdisciplinary and comparative, and the results obtained are used in educational activities. It is also of great importance for CEDMO researchers to be able to collaborate with communities involved in countering disinformation, especially fact-checker journalists.

The activities of the Central European Digital Media Observatory can be divided into two phases. The first, implemented in the years 2021–2024 and co-funded by the European Union (action no. 2020-EU-IA-0267) and by the Polish Ministry of Education and Science under the programme of the Minister of Science and Higher Education entitled "PMW" in the years 2021–2024 (contract no. 5213/CEF/2021/2) and the second, implemented in the years 2024–2026 as CEDMO 2.0 funded by the European Union (identification no 101158609). In this second phase of its activities, CEDMO is continuing its work from previous years and introducing new initiatives (also of a research nature) in a consortium expanded to 13 partners.

SWPS University, also supporting this publication, has been at the forefront of a package of research activities since the beginning of the project. As part of these activities, we aim to gain in-depth scientific knowledge on disinformation, as well as to develop a research model to continue and develop research on this important and complex phenomenon in the future.

Views and opinions expressed are however those of the authors only and do not necessarily reflect those of the European Union. Neither the European Union nor the granting authority can be held responsible for them.

Funded by
the European Union

Contents

Notes on Contributors

Filip Cyuńczyk is a lawyer, sociologist of law, and historian of law. He is assistant professor at the SWPS University Institute of Law. His scholarly interests include the sociology of law, with a particular focus on the study of constitutionalism, the relationship between law and group identity, the legal politics of conflict resolution, the legal regulation of collective social memories and the study of legal education. He is a visiting lecturer at the Riga Graduate School of Law and the European Humanities University in Vilnius.

Lucia Furtáková is a PhD. candidate at the Faculty of Mass Media Communication of the University of Ss. Cyril and Methodius in Trnava. Her research activities focus on digital literacy and competence, radio and television production.

Beata Gavurova is a Professor in the field of Management and Economics of Enterprises. Her research activities are focused at the support of creation of health, social, and environmental policies, sustainable development processes, solving issues of the regional disparities and discrepancies at the national as well as international level.

L'uboš Greguš works at the Faculty of Mass Media Communication of the University of Ss. Cyril and Methodius in Trnava. His scholarly expertise involves electronic media reporting, with an emphasis on television, the online environment and its audience.

Nik Hynek is Professor at the Faculty of Social Sciences of Charles University in Prague. In the past, he worked within the Czech Ministry of Foreign Affairs as a researcher and established a research centre on security. He was a visiting research scholar at the Saltzman Institute of War and Peace Studies at the Columbia University in New York; The London School of Economics and Political Science; Australian National University; PRIO; Carleton University in Ottawa; and Ritsumeikan University in Kyoto.

Viliam Kovac is a researcher focusing on the application of quantitative methods and explanatory research in economics. He currently works at the Faculty of Mining, Ecology, Process Control and Geotechnologies of the Technical University in Košice. In his scientific and research activities, he focuses mainly on the environment, tourism, health aspects, especially in the economic context.

Peter Krajčovič works at the Faculty of Mass Media Communication of the University of Ss. Cyril and Methodius in Trnava as assistant professor. His scholarly expertise involves media research, media communication and disinformation.

Jakub Kuś is a psychologist. He works on the psychology of new technologies, with a particular focus on research on the dispersion of responsibility on the Internet and distortions in the reception of digital content. Recently, he has been focusing on research in the psychology of disinformation, especially digital disinformation.

Lenka Labudová works at the Faculty of Mass Media Communication of the University of Ss. Cyril and Methodius in Trnava. Her scholarly expertise involves regional marketing and marketing communication.

Veronika Macková is an academic researcher at the Faculty of Social Sciences, Charles University in Prague. Previously, she held a position as a journalist at Czech Television, which is why she mainly teaches practical journalism-oriented courses. Her areas of research interest include sports journalism and artificial intelligence journalism.

Vaclav Moravec is the lead coordinator of the Central European Digital Media Observatory (CEDMO). He works as an Assistant Professor at the

Institute of Communication Studies and Journalism, Faculty of Social Sciences, Charles University and at the Department of Production, FAMU, Prague. He specializes in the transformation of audiovisual media, journalistic ethics, automated journalism, and journalism with artificial intelligence.

Jana Radošinská reflects on media culture, mainstream film production, media industries and entertainment forms. She has published several books on globalised media industries (film, television, music, digital games) and their synergies. Her scholarly expertise involves superhero films, entertainment hybrids present in audiovisual media, and current perspectives on understanding media audiences.

Martin Solík works at the Faculty of Mass Media Communication of the University of Ss. Cyril and Methodius in Trnava as assistant professor. His scholarly expertise involves contemporary moral, social and political philosophy with a focus on the concept of recognition in local and transnational contexts.

Karina Stasiuk-Krajewska is associate professor at SWPS University. She is the coordinator of Central European Digital Media Observatory in Poland. Scientifically she deals with issues of ethics, professionalization and disinformation as well as quantitative and qualitative discourse analysis. She is a member of EDMO's Group of Experts on Structural Indicators for the Code of Practice on Disinformation and IFCN expert.

Kateřina Turková is a researcher and associate lecturer at the Institute of Communication Studies and Journalism, Charles University, Prague, Czech Republic. In her academic career, she primarily focuses on issues associated with sports, social media, and quantitative research methods.

Ján Višňovský works at the Faculty of Mass Media Communication at the University of Ss. Cyril and Methodius in Trnava as a university professor. His scholarly expertise involves current issues of journalism in the context of the development of information and communication technologies, online journalism and the history of television in English-speaking countries

Michał Wenzel is associate professor at SWPS University. His areas of interest are research methods and sociological aspects of the media. In the past, he worked at CBOS Public Opinion Research Centre and the University of Oxford. He completed research stays at the University of Michigan and the Max Planck Institute in Cologne.

Klára Zubková is a PhD. candidate at the Faculty of Mass Media Communication of the University of Ss. Cyril and Methodius in Trnava, focusing on artificial intelligence, disinformation, critical thinking, and educational games.

List of Figures

List of Charts

List of Diagrams

List of Tables

CHAPTER 1

Introduction

Karina Stasiuk-Krajewska and Michał Wenzel

Disinformation is undoubtedly one of the major problems of the modern world. As such, it has become a subject of interest not only for professionals (journalists, fact-checkers, media literacy experts) and researchers, but also for politicians and the general public.

This interest in disinformation, and the actual expansion of the phenomenon of disinformation, stem from several factors. First, in recent years we have witnessed significant upheavals in the world, which have shaken our faith, as residents of the so-called "first world," in its order and guarantee of security. The pandemic, armed conflicts, the migrations of large groups of people, and the general polarization of politics and society—all of this creates fertile ground for disinformation, which thrives on these kinds of sentiments. As the sense of security and faith in world order erodes, trust in those who should guarantee this security and ensure world order—politicians, representatives of international institutions, scientific and moral authorities—diminishes.

K. Stasiuk-Krajewska (✉)
Faculty od Law and Communication, SWPS University, Wrocław, Poland
e-mail: kstasiuk-krajewska@swps.edu.pl

M. Wenzel
Dept. of Social Sciences, SWPS University, Warsaw, Poland
e-mail: mwenzel@swps.edu.pl

M. Wenzel, K. Stasiuk-Krajewska (eds.), *Disinformation in Central Europe*, https://doi.org/10.1007/978-3-031-76871-2_1

Social phenomena of a broader nature are accompanied by specific changes in media systems, which constitute the primary environment for the production of information, and therefore disinformation as well. We are observing, first, a decline in trust in traditional media, moreover, a general decline in interest in the content generated by these media (Digital News Report, 2022). Second, for a number of reasons (commercialization, polarization, sensationalism), the quality of professional journalism is actually declining, and it has to adapt, on the one hand, to the expectations of its audience, and on the other, to the demands of the market and editorial boards. And finally, last but not least, social media have become a powerful force of communication, which due to its specific structural and technological features provides an excellent environment for creating and spreading disinformation (Khan et al., 2022; Bennet & Livingston, 2021).

The issue of technological advancements in artificial intelligence, a recent subject of intense debate, raises significant concerns. This issue is crucial, and the consequences of producing fictitious messages, indistinguishable from true ones by non-experts, could lead to a complete loss of trust in any sources of knowledge or information in the future. More specifically, it could result in the complete equalization of entirely unreliable sources with those that are actually trustworthy.

Given this, it is no surprise that the growing interest in disinformation is also evident among researchers, both from the fields of social sciences and humanities, and those involved in modeling and designing processes involving artificial intelligence. There is a general consensus that disinformation is a complex, dynamic, and global phenomenon, which poses significant challenges for its scientific study.

However, when studying disinformation, we encounter a more fundamental and likely more difficult problem: defining the phenomenon itself. In the classic definition, disinformation (as opposed to misinformation and malinformation) is defined as the intentional dissemination of false or inaccurate information with the goal of misleading the audience (Wardle & Derakhshan, 2017; Tandoc et al., 2018).

However, such a definition is quite broad. In this context, there is a tendency to apply the category of disinformation to all lies and manipulations, and even techniques of social influence that circulate in the public sphere. To define disinformation, however, it is necessary to identify its characteristic features, linked to the mechanics of political or social processes that we observe today or, what in my opinion is more promising and in no way contradictory to the former, the mechanics of mediated communication processes.

The attempt to refine this definition led to the introduction of the category of FIMI (Foreign Information Manipulation & Interference). In this view, FIMI is an activity that is: Harmful (for democracies, societies, values, security etc.); not illegal (not covered by other legal instruments, located in a "grey zone"); manipulative (which includes manipulation of content); intentional (it is not misinformation, but pursues a specific goal and is organized/financed); coordinated (it uses different parts of the "ecosystem" and is not "organic").

The formulation of this category has become the basis for proposing a broad model or research program on FIMI phenomenon, which undoubtedly has a high value. However, the definition itself is still not unambiguous or precise.

The term "fake news" is often used synonymously with "disinformation", however, it's important to clarify its meaning and scope (Shu & Liu, 2019; Molina et al., 2021; Lazer et al., 2018). It should not be used as a general term encompassing satire, parody, public relations, and advertising, but rather as a specific type of media phenomenon, which can be analyzed through the lens of sender, receiver, and, most importantly, communication structure. In this approach, "fake news" would be a deliberately misleading message that imitates real news or journalistic information. In this narrow sense, "fake news" is a specific type of media phenomenon that presents a false representation of reality, about which it can be definitively determined that it is indeed false (this is how fact-checkers understand the category of "fake news," clearly labeling verified fake news as true or false).

This proposed definition of "fake news" is a significant step towards defining the scope of research, but it has its drawbacks. First, and rather obviously, a considerable amount of actions undertaken by "bad actors" with malicious intent do not have the structure of "fake news". Therefore, a significant portion of these actions remains outside this defined research area. On the other hand, it is worth noting that a broader perspective is possible. One could argue that disinformation includes not only texts imitating real news ("fake news" in the narrow sense) but also all media messages that, according to the sender's intention or belief (including institutional sources) and the receiver's expectations, fulfill functions traditionally assigned to journalism or media, including primarily the information function. They draw upon seemingly authentic values like truth and rationality, which constitute the relationship between sender and receiver in the realm of journalism (institutional media). This approach would expand the category of disinformation to include textual creations that are formally similar to documentaries, reports, or rational opinions. On the other hand, it would allow for the exclusion of satire, humor, advertising, gossip, etc., because these are not discursive practices that are embedded in the appropriate communication framework (in quasi-journalism discourse) for disinformation.

In this proposed approach, the intention of the sender of disinformation would not be to mislead about the truth of the world, but rather to establish a false communicative relationship with the receiver—a false relationship because it relies on discursive and professional practices (e.g., verifying information from multiple sources) that are not actually being implemented. Disinformation presents itself as truthful, credible, socially responsible (acting for the common good) and fulfilling the receiver's right to information. This is the same set of values that constitutes the semantic chain of journalism. The "fakeness" of these values stems not necessarily from the sender's intention (though that is possible), but rather from the fact that the resulting texts do not meet the basic requirements of professional discursive practices in journalism, but are nevertheless presented to the receiver as such. Disinformation appropriates the functions and discursive features of journalism, or more precisely, these features are assigned to it by the recipients of disinformation through their interpretative practices. Effective disinformation is read, shared, forwarded, commented on, etc., as truth about the world.

It is also important to remember about the psychological (Greifeneder et al., 2021) and social consequences of the spread of disinformation (O'Shaughnessy, 2020), which include: creating, maintaining, and strengthening political or social divisions; sowing confusion with the intention of creating a sense that nothing is credible; and finally, raising doubts, which is an effective tool of disinformation, because even credible facts can rarely be proven absolutely (e.g., in the case of scientific research).

It is clear that disinformation strikes at the heart of democratic values, at the sense of rationality in the world, respect for others, the conviction that we can come to an understanding and that cooperation is better than conflict. Furthermore, disinformation undermines trust in truth, journalistic truth (as mentioned above), but also scientific truth. This is particularly important because truth (along with reason) is a constitutive value of modern societies.

Considering all the observations and concerns raised above, we would like to propose a way of understanding disinformation that allows us to view it from a broad theoretical and methodological perspective, taking into account its structural features (linguistic and visual) as well as the issues related to its dissemination and reception. We consider these latter issues from several perspectives—as psychological, social, political, and legal mechanisms.

The contents of this volume can be divided into two sections. The first part is composed of theoretical approaches to three different aspects of disinformation: its structure, reception and communication. The second section contains quantitative and qualitative empirical studies. Finally, we complete the volume with a reflection on legal consequences of combating disinformation.

Chapter 2 is an introduction to the issues of disinformation in the context of discursively oriented theory and analysis of media genres. The purpose of the chapter is to introduce the specifics of the genre as it is understood in media studies, especially in the context of discourse theory. It offers theoretical solutions from the field of linguistically oriented media studies that can be applied to the analysis of fake news as a genre. Such an analysis provides an insight into the structural features of disinformation.

Chapter 3 explores the psychological dimensions of disinformation. The author explains what makes individuals susceptible to disinformation, highlighting the role of emotional arousal, information deficits, and digital environments in amplifying vulnerability. The text discusses the impact of disinformation on society, including its effects on the economy and politics. The authors emphasizes the importance of critical thinking, source verification and emotional regulation as key psychological tools in combating disinformation.

Chapter 4 contains the description of mechanisms of disinformation. The authors discuss the spread of propaganda, fake news, and disinformation in Central Europe. The goal of the chapter is also to reveal the strategies that have been used for the dissemination of these messages amongst the societies in the region. It contains an overview of the misleading narratives that accompanied the Covid-19 pandemic and war in Ukraine in the Central European countries. In both cases, the essential role of online and social media in the spread of disinformation was confirmed.

Chapters 5 and 6 are empirical studies conducted in the qualitative paradigm. Chapter 5 presents the results of research on a corpus of fake news in the Polish, Czech and Slovak language. The research was conducted using assumptions and methodology inspired by linguistic language analysis and van Dijk's theory of discourse and media genre. The narratives and discursive elements that produce the legitimacy of fake news and their subject matter are analysed. Chapter 6, on the other hand, focuses on the visual aspects of fake news in the context of news values theory. Both studies are comparative. A transnational model of fake news and its variants for individual countries is proposed, referring to both structural and semiotic features.

Chapter 7 is an empirical study in the quantitative paradigm. The twin crises of Covid-19 and the war in Ukraine are treated as a 'polycrisis' and the interrelation between the attitudes to both are examined based in the results of mass surveys conducted in Poland, the Czech Republic and Slovakia. The attitudes are linked. The scepticism towards the official state policy to Covid-19 mutated into the negative attitude to policy towards the war, and hence to a pro-Russian attitude. The attitude to Covid-19 is uniform across the three societies, whereas the attitude to the war in Ukraine differentiates them. In Poland the pro-Russian narratives are met with the weakest response, while Slovaks are the most susceptible to this type of influence.

In Chap. 8 the authors investigate the perceived importance of regional partnerships and the extent of perceived electoral interference by major powers in the Czech Republic, Poland, and Slovakia. Grounded in Bauman's liquid modernity, which emphasizes fluidity, instability, and fragmentation in modern societies, the study explores how socio-demographic characteristics and political party preferences influence these perceptions. Regression analysis reveals that age and education significantly affect perceived partnership importance. The study also identifies political parties with high and low odds of perceived electoral interference by major powers such as China, Russia, Germany, the United States and the European Union.

Chapter 9 completes the volume. The author explores the relationship between law and politics in the context of disinformation, particularly its effective counteraction within the system of constitutional liberal democracy. His objective is to identify the key challenges that may arise when applying normative regulations to counter disinformation phenomena. It may impossible to draw a clear line between cynical disinformation efforts by participants in public discourse (within a specific political community) and minor violations of its principles. The author addresses scenarios where the law is used—as a tool by political actors—to combat external disinformation. This publication is part of an international project co-funded by the European Union (action no. 2020-EU-IA-0267) and by the Polish Ministry of Education and Science under the program of the Minister of Science and Higher Education entitled "PMW" in the years 2021–2024 (contract no. 5213/CEF/2021/2). Views and opinions expressed are however those of the authors only and do not necessarily reflect those of the European Union. Neither the European Union nor the granting authority can be held responsible for them.

References

Bennet, L., & Livingston, S. (Eds.). (2021). *The disinformation age. Politics, technology and disruptive communication in the United States.* Cambridge University Press.

Digital News Report. (2022). Reuters Institute. https://reutersinstitute.politics.ox.ac.uk/digital-news-report/2022

Greifeneder, R., Jaffe, M., Newman, E., & Schwarz, N. (2021). *The psychology of fake news: Accepting, sharing, and correcting misinformation.* Routledge.

Khan, A., Brohman, K., & Addas, S. (2022). The anatomy of 'fake news': Studying false messages as digital objects. *Journal of Information Technology, 37*(2), 122–143.

Lazer, D. M. J., et al. (2018). The science of fake news. *Science, 359*, 1094–1096.

Molina, M. D., Sundar, S. S., Le, T., & Lee, D. (2021). "Fake news" is not simply false information: A concept explication and taxonomy of online content. *American Behavioral Scientist, 65*(2), 180–212.

O'Shaughnessy, N. (2020). From disinformation to fake news: Forward into the past. In P. Baines, N. O'Shaughnessy, & N. Snow (Eds.), *The Sage handbook of propaganda.* SAGE.

Shu, K., & Liu, H. (Eds.). (2019). *Detecting fake news on social media.* Morgan & Claypool Publishers.

Tandoc, E. C., Lim, Z. W., & Ling, R. D. (2018). Fake news: A typology of scholarly definitions. *Digital Journalism, 6*(2), 137–153.

Wardle, C., & Derakhshan, H. (2017). *Information disorder: Toward an interdisciplinary framework for research and policymaking* (Vol. 27, pp. 1–107). Council of Europe.

CHAPTER 2

Disinformation (Fake News) as a Text: Genre and Discursive Perspective

Karina Stasiuk-Krajewska

[...] every thing and every idea that we 'know' is known through sign-systems. And sign-systems have properties which 'speak us' even as we create meaningful utterances. Hence, for news, what makes it meaningful is not the world it reports but the sign-system in which it is encoded.

—Hartley *(2013: 33)*

2.1 Introduction

There is no doubt that the category of disinformation is an extremely imprecise and broad one, which includes phenomena that are diverse, both from the formal point of view and taking into account the social and communication contexts in which they occur. The situation is further complicated by the fact of unclear boundaries between disinformation and concepts that in the literature on the subject are regarded as

K. Stasiuk-Krajewska (✉)
Faculty od Law and Communication, SWPS University, Wrocław, Poland
e-mail: kstasiuk-krajewska@swps.edu.pl

M. Wenzel, K. Stasiuk-Krajewska (eds.), *Disinformation in Central Europe*, https://doi.org/10.1007/978-3-031-76871-2_2

related—malinformation and misinformation (Wardle & Derakhshan, 2017). An attempt to clarify the scope and define the specifics of phenomena that are associated both in social awareness and in scientific literature with the category of disinformation is undoubtedly the introduction of the FIMI category—Foreign Information Manipulation and Interference (Buvarp, 2023). The introduction of this concept is accompanied by a rather detailed characterisation of it and, perhaps more importantly, a proposal for an overall research model that can (and should) be applied in the context of research into this phenomenon. However, a closer look at the definition of FIMI may raise significant doubts (for example, whether the restriction of disinformation activities to foreign operations is justified and whether this type of area can be distinguished at all). On the other hand, the proposed research model still remains at a very high level of generality, and the operationalisations of the model that appear in various studies are of a very diverse and uncorrelated nature.

Therefore, without denying the need for research into the broader phenomenon of disinformation or FIMI, I would like to focus in this chapter on a small part of this phenomenon—fake news. Clearly, such a decision entails significant limitations, but it does at least allow for some sort of structured study of disinformation to begin. First and foremost, the category of fake news is relatively precisely defined (for the standards of social sciences and the field of disinformation) (Egelhofer & Lecheler, 2019). Fake news can be considered to be a specific media genre, structurally and semantically imitating news (Lazer et al., 2018), which is characterised by the fact that it presents a completely or partially false picture of reality, which can be unambiguously stated to be such. This is how the category of fake news is understood by fact-checkers, who label the fake news they verify with the unambiguous categories of true or false. Thus, fake news should be understood as an imitation of news. Although it has a structure and semantics similar to real news, it differs from it in important ways: 'Specifically, real news stories are stories that are known to be true and from well-trusted news sources. Fake news stories are stories that are known to be false and from well-known fake news websites that are intentionally trying to spread misinformation' (Horne & Adali, 2017). Let us note, however, and this will be an important point of discussion in this chapter, that fake news is considered true by its audience, just as the sources to which it refers are treated as credible.

Some approaches add the intention to deceive to the two previously indicated distinctive features of fake news, which, as will be discussed below, has a deep justification in the theory of genre, but is sometimes difficult to state unequivocally, especially if we work only on textual material. It also seems (although, it is difficult to determine on the basis of an analysis of the formal and semantic structures of the texts) that not all authors of fake news actually realise this intention to mislead. Obviously, the definition above (fake news as imitation news) can only be accepted if one rejects, which I do in this chapter, the classifications of fake news defined in the literature as classifications 'in a broader sense', which, apart from false or manipulated content, include, among others: satire, parody, image manipulation, public relations, advertising, propaganda, rumour, conspiracy theories, or urban legends; and even commentary, polarised content, persuasive information, and citizen journalism (Khan et al., 2022; Shu & Liu, 2022).

The phenomenon of disinformation, and therefore also the phenomenon of fake news, which is an integral part of it, undoubtedly has a medial character in the sense that it functions in and through media, is mediated by them and in its social existence is dependent on them. Although disinformation, biased information, advertising or information used to discredit certain values or beliefs have always been part of socjety (Cortada & Aspray, 2019), the current development and specificity of this type of phenomenon is undoubtedly linked to the evolution of media systems and structures, especially the development of social media (Baptista & Gradim, 2020). Functioning in the media, disinformation takes the form of a media message, text, in a broad multimodal sense. Studying the structures, semantics and functions of this text is one of the important directions of research on disinformation, which also has significant practical significance, as it enables the creation of automated tools for detecting fake news.

There are relatively few studies on fake news as texts, dominated by those which refer to the assumptions of corpus linguistics, based on relatively large corpora of texts, but usually limited to the presentation of grammatical structures or lexical elements, to a lesser extent referring to semantics and to a very limited extent, to discourse, understood as structure and semantics from above the sentence level (Majid Akhtar et al., 2023; Horne & Adali, 2017). Other approaches, on the contrary, are oriented mainly not to analysis of the structures present in fake news, but to analysis of the use of semantic resources (textual, visual) or semantic procedures (e.g. framing) for persuasive or manipulative purposes (Juez & Mackenzie, 2019), although there are also some that refer directly to the category of genre (Egelhofer & Lecheler, 2019). All of them carry significant cognitive value, and the theoretical frameworks and research results presented in this and the following chapters on language and visuality are part of this way of understanding fake news, aiming to outline a theoretical framework, but also to present the implementation of certain research concepts that can increase knowledge of fake news treated as texts. My aim in this chapter is, therefore, is to outline preliminary theoretical and analytical areas within which fake news could be described as a genre of pesudojurnalistic disinformation (Egelhofer & Lecheler, 2019), areas that should certainly be subject to further clarification and probably also, to some extent at least, reformulation in the course of further research. The decision to adopt a discursive perspective of genre analysis in the studies on fake news stems from the fact that this approach is not only one of the most important in the study of media texts in general, but also becomes the dominant approach in the analysis of textual manifestations of disinformation (Maci et al., 2024).

2.2 Genre: The Limits of the Concept

Although there are many, not necessarily coherent and rather strongly differing, definitions of genre, there are a few points on which the representatives of different approaches and paradigms agree.

The first and foremost assumption is that the genre must be situated in a certain way, rooted in a situation external to itself. This means that it is actualised in a certain social situation, a specific space or within a specific community. It is difficult, in a scientific sense, to speak of a genre that is not subject to processes of social actualisation, that is not (was not) realised in any way.

Secondly, a genre must realise certain rules that relate to its internal structure and at the same time allow the participants in communication (in the broad social sense or in the limited sense of a specific communicative community) to distinguish it from other genres. The above assumption does not mean, of course, that the structure of the genre does not change over time—but each time a new (modified) genre has to be recognised as such (as distinct from what existed before) by a given communicative community (Yates & Orlikowski, 1992). The possible recognition of the emergence of new genres (for example, on the internet) requires not only the emergence of new structures, but also their recognition as different and distinctive by the audience. Thus, the question of news on the web is in fact a question of whether audiences recognise its distinctiveness from news present in, for example, the printed press. Similarly, the ultimate recognition of the genre distinctiveness of fake news requires confirmation of the existence of this distinctiveness in the minds of the audience. I will return to this issue later in this chapter.

Generally speaking, the genre must therefore be characterised by a certain integrity of both an internal and an external nature. Integrity at the first level is built by the relations taking place within the textual universe: relations within a given text (lexical, rhetorical-grammatical, and discursive); as well as contextual and intertextual relations. Integrity at the second level goes beyond the text universe towards the social: it is related to discursive procedures and practices as well as the culture of a given community, e.g. professional (Bhatia, 2014a). Thus, a text that we analyse as representative of a given genre must 'have verisimilitude in relations to two chief kinds of norms. The first is what we call the rules of the genre: for the work to be said to have verisimilitude it must conform to these rules. [...] But there also exists another verisimilitude, which has been taken even more frequently for a relation with reality. Aristotle, however, has already percived that this verisimilitude is not a relation between the discource and its referent (the relations of truth), but between the discourse and what readers believe it is true' (Neale, 1990: 47).

Considering news as a genre, we can expect at the first level of integrity a typical textual structure (as will be discussed below), a specific contextual embedding in relation to other texts (e.g. data sources or reports) and specific interrelationships with other texts in the genre field (with other news for example). At the second level of integrity, procedures for the production of reliable information, realised as editorial practices (e.g. related to checking sources) and the existence of a coherent and socially present professional discursive journalistic community are important.

All these conditions are fulfilled in the case of news. Could their fulfilment also be demonstrated in the case of fake news? If we were to accept the hypothesis of fake news as a constituted genre, we would have to be able to demonstrate not only a relatively stable internal structure or relations with other texts (which seems possible), but also a relatively unambiguous location of the 'fake news genre' in what is social—in the world of procedures, organisational (group) culture or stabilised communication practices. It would therefore be necessary to demonstrate that there are specific procedures and discursive practices for the production of fake news and a kind of 'organisational culture' of groups of senders and receivers of this type of message.

The conditions for distinguishing a particular genre, however, are not only intrinsic formal issues or discursive practices of communities external to the text, but also a specific communicative purpose, or, to put it another way, a specific function in communicative processes.

The unquestionable (though not the only) function of news would certainly be to inform. This function is essentially unquestionable, as it is clearly visible both in the internal structure of the text and in the broadcasting and reception practices, and finally in the relatively consistent social expectations shared by both the senders (journalists) and the receivers (e.g. press readers). In the case of fake news, on the other hand, the question of function is much more complicated. In many cases, it is constructed as informative (both in the structure of the text and in the intentions of the sender and the expectations of the audience). However, the definition of fake news indicates not only that it is misleading (which would not necessarily contradict the informational function, as this function is not objective, but rather involves certain textual structures and the text's relationship with the sender and the recipient), but also the sender's intentions, involving misleading. However, the issue is not clear-cut—the sender may intentionally mislead the receiver and, at the same time, aim to have the receiver treat the message as true and therefore attribute an informative function to it. This appears to be the main modus operandi of fake news, although the issue certainly requires further research and analysis.

The genre is thus a recognizable communicative event characterized by a set of communicative purpose(s) identified by the members of the community in which it regularly occurs. This kind of event is highly structured and conventionalized in the intent, positioning, form and functions (Bhatia, 2013). Viewing genres through the prism of their dominant function is also reflected in their classification, which in classical terms includes (Longacre, 1974): the narrative genres (with a describing function), the procedural genres (with an instructing function), the expository genres (with an entertaining function), and the horatory genres (with a persuading function). In the context of this classification, news obviously belongs to the category of narrative genres, as does fake news (let us note that the describing function does not determine whether a story is true or false—narrative genres also include, for example, fairy tales or myths).

Considering the question of the function of genre is particularly important in the context of discourse-oriented theory and analysis, as it provides an opportunity to ask not only: how? but also: why? This question (and the attempt to answer it) can be considered the essence of the paradigm of discourse analysis in the social sciences in general. It is also fundamental in the context of the analysis of fake news. For we are not only interested in answering the question of what their structure or patterns of dissemination are, nor even in answering the question of their impact on media audiences or social harm. These questions are important, but the more cognitively relevant issues relate to the category of function, which is in fact the other side of the question: why? Why is fake news created? Why do recipients treat it as a reliable source of knowledge about the world? Why is it structured the way it is and what is the function of this structure? These are just a few of the questions worth asking in this context. However, in line with the assumptions made above about the constitutive features of the genre, it is also necessary to reflect on the external and internal coherence of the fake news genre, as well as on the typical communicative situations within which the texts representing it appear and circulate.

2.3 Genre in the Context of Language Research: Towards a Definition

Genre is, by its very nature, a communicative phenomenon, 'a type of text' (assuming a broad understanding of the category of text as a meaningful syntagm) or 'communicative event'. Hence, its study will always be a study of the signs that constitute this text or communicative event. To put it differently—it is difficult to imagine the analysis of any genre without the analysis of its sign/semantic structure. This is why the study of genre is characterised by the same perspectives that are present in the study of texts: primarily linguistic, sociological and psychological (Bhatia, 2013). The first of these, strictly linguistic, focuses on the linguistic description of texts, their lexical, grammatical or discursive features. This approach looks for characteristics specific to the texts in question, and then treats these characteristics as constitutive of the genre (e.g. news or fake news). The sociological perspective, called sociosemiotics (Kress, 2009), focuses on the analysis of how genres relate to social reality—how they define, organise or communicate it. This approach emphasises above all that a text is not a phenomenon the meaning of which functions in isolation from social extra-textual reality; that—when analysing genres—social roles or processes within social groups must be taken into account. A text, pursuing certain genre patterns, is always situated in a certain social context, a social situation, which not only remains in a dynamic relationship with its internal structure, but also determines the functions it performs. Such a perspective therefore makes it possible to seek answers to the question of why a given text in a given social situation adopts such a genre structure and not another. In this view, the study of fake news would mean, for example, addressing the typical social situations (e.g. using social media) within which fake news circulates, or the functions it performs in contemporary societies. The psychological (or more strictly psycholinguistic) perspective refers primarily to the cognitive structures, and in particular—the strategies and tactics that the author of the text employs to achieve his/her goals while remaining within the socially accepted genre framework. The essence here is the use of conventional textual principles in a specific socio-cultural context, also taking into account the expectations of the audience. In the case of the media, both broadcasters and audiences share a common understanding of what the news serves; they have a common understanding that it has a primarily informative function. Consequently, journalists employ certain communicative tactics that make their texts credible as news in its informational function, while audiences decode texts within their assigned function (as informational). These processes are cognitive in nature and reflect the strategies that the representatives of the discursive communities (for example, professional communities) use to achieve their communicative goals. The use of these strategies and tactics requires the subject to refer to specific knowledge that is professional in nature and, as such, not available to actors from outside the professional community. Thus, the use of these tactics constitutes the social actor as a member of a given discursive community.

From such a perspective, one could ask, for example, what strategies and tactics are used by the creators of fake news and what communication goals they pursue in this way. The question of the relation of fake news to conspiracy theories, which can be regarded as a certain type of 'world knowledge', could also be of interest.

The evolution of the study of text is also an evolution of the study of genre as a textual phenomenon. Nowadays, in analytical practice, the perspectives presented above interpenetrate each other, which is also perfectly evident in the dominant conceptions of genre.

Thus, given the characteristics of genres mentioned earlier (their social location, external and internal integrity and, finally, functionality) as well as the fact that they are essentially textual (semantic) phenomena, they should be defined as systems of codes, conventions and social expectations (Feuer, 1992). Thus, genre can be treated as a hybrid phenomena that 'mediates between the industry and the audience' (Erjavec & Poler Kovačič, 2009: 149). This complexity and multifaceted nature of the genre category can be seen perfectly in the context of film production, where 'genres do not consist only of films: they consist also, and equally, of specific systems of expectations and hypothesis which spectators bring with them to the Cinema, and which interact with the films themselves during the course of the viewing process' (Neale, 1990: 46). Genre thus can be understood in several quite convergent ways, as:

- 'a class of communicative events that share a recognizable communicative purpose, that exhibit a schematic structure supporting the achievement of that purpose, and which show similarities in form, style, content, structure and intended audience' (Swales, 1990: 58);
- 'staged, goal-oriented, social processes' (Martin et al., 1987: 59);
- 'a conventional category of discourse-based in largescale typification of rhetorical action' (Miller, 1984: 163);
- 'a distinctive type of communuication action [...], characcterized by a socially recognized communicative purpose and common aspect of form' (Orlikowski & Yates, 1994: 543).

In view of the complex characteristics of genre (as a textual phenomenon, but also as a social practice), it is of course possible to adopt three perspectives: one focused on the text (which is how genre has been analysed by formal linguists), one focused on social action, and one perspective combining both. Indeed, in some approaches to genres, located within the broader framework of discourse theory (or even more broadly, linguistic research), one can see a shift from the study of the text as such and the meanings it generates towards the study of the social action that the actualisation of genres entails. Genres are here seen as '(inter)actional formats that can accommodate an ever widening range of discourses, and as strategies for achieving objectives, rather than, for instance, as resources for negotiation' (Gruber, 2008: 359). In this approach, close to linguistic pragmatics, genres are treated as elements of interaction which, by producing a specific relation between subjects and performing specific functions, constitute the system's response to frequently recurring communicative situations. Such an approach, while still remaining in the field of discursive analysis, refers explicitly to the Goffmanian tradition. Goffman analyses how participants in interactions produce and execute communicative strategies and scenarios, and how they manage their communication in order to achieve the desired outcomes. In the context of the above insights, Goffman develops both the interactional ritual model and the strategic interaction model. These assumptions are also the basis for Goffman's formulation of the concept of *frame* analysis, which treats frames as specific communicative rules that establish the basis of interaction, and thus, on the one hand, build up the recipient's expectations of the text, and on the other hand, limit the freedom of the sender, who (if he/she wants his communicative behaviour to fulfil a certain function) has to adapt to these frames (Goffman, 1974). The concepts of *frame* and *framing* have gained considerable popularity in media studies, where they are used primarily in the context of the study of genre media structures, such as journalistic news (De Vreese, 2005; Pan & Kosicki, 1993; D'Angelo & Kuypers, 2009).

In the context of researching fake news, it is first and foremost necessary to adequately formulate a definition of the genre—so that, on the one hand, it takes into account the latest research on the subject and is sufficiently accurate, and, on the other hand, it can be operationalised in relation to a phenomenon as specific as fake news.

2.4 Genre as Model

The special role of the study of textual structures of fake news derives from yet another feature of genre, that is often overlooked in media studies analyses. Whichever definition of genre one adopts, it should be remembered that genres, epistemologically speaking, are not real existing phenomena. Genres should therefore only be considered as models or structures that are realised, to a greater or lesser extent, in communicative practice. Moreover, genre integrity is not something static, given once and for all. On the contrary, it is constantly challenged, negotiated and changed in everyday communicative practice.

A genre is thus an abstract creation (a model, a pattern), which nevertheless has various concrete realisations in the form of expression, and which also exists as a set of conventions that tell the members of a particular communicative community what shape to give to specific interactions (Wojtak, genres). If genre is a formal pattern, the text is a realistically existing, more or less accurate, realisation of that pattern. Sometimes, in this context, one speaks of a genre pattern (Fras, 2013: 11), i.e. a potentially real text realising all genre determinants (archetext). However, in the practice of genre research, especially journalistic or media genres, finding such archetexts does not seem possible.

If a genre is understood as a pattern, it must be assumed that there are more or less exact realisations of it. These realisations function within the genre field, i.e. the collection of utterances realising the genre. 'The central area of the field is occupied by the canonical pattern, the sphere close to the centre by alternation patterns, and in the peripheral zones, i.e. furthest from the centre and close to the boundaries of fields typical of other genres, adaptation patterns are situated. This is also the area of species interbreeding and the formation of hybrids' (Wojtak, 2004: 18). Such an approach to genre—as a field, and not just a canonical pattern—allows us to explain the diversity of its realisation in practice, to show the evolution and dynamics of change, as well as—what seems particularly important nowadays—the relations with the fields of other genres (e.g. with the field of entertainment genres, resulting in the phenomenon called *infotainment*). The degree to which a given text implements genre conventions also varies, of course. In the field, we can distinguish between codified genres (e.g. in the form of existing official documents), fixed genres (in written form or in social memory), conventionalized genres (here an element may be subject to exchange or transformation), but also—and here we enter the fringes of the aforementioned genre field—unconventionalized genres (characterized by one important feature) and new genres (i.e. fluid genres, those that have not managed to develop the characteristics of a pattern) (Wilkoń, 2002: 201).

The realisation of the ontological status of genre as a pattern and the recognition of the differentiation of genres in the field (depending on the extent to which they realise the pattern) opens up new perspectives for the study of fake news as a genre. It allows us, first of all, to accept the diversity of its textual realisations and, consequently, to recognise that this diversity does not preclude the search for a generic description of the phenomenon. Acknowledging the diverse relations between texts functioning in the genre field also opens up the possibility of asking questions about the interrelationship of texts representing, to a greater or lesser extent, structural and functional features of fake news, and thus opens up the field for reflection on other textual manifestations of disinformation.

A different attempt to organise the genre field is the notion of colony of genre. 'Genre colonies represent groupings of closely related genres serving broadly similar communicative purposes, but not necessarily all the communicative purposes in cases where they serve more than one. In addition they may not share a number of other features of context' (Bhatia, 2014: 67). An example of a genre colony may be promotional genres, in the field of news studies—genres whose function (projecting structure) is to report facts. In this context, it is perhaps not appropriate to speak of fake news as a genre, but rather a genre colony that formally mimics news genres, while functionally (in an objective sense) disinforming.

The category 'genre colony', like the category of the field, is an important attempt to bring order to the somewhat chaotic field of genre theory. 'Firstly, it brings a degree of versatility to genre identification and description, in that it allows genres to be viewed at different levels of generalisation, making it possible to posit principled relationships between super or macro-genres, genres, and sub-genres. It also makes it possible to relate these subcategories to features of context. The concept itself incorporates two related meanings. Firstly, it represents a grouping of closely related genres, which to a large extent share their individual communicative purposes, although most of them will be different in a number of other respects, such as their disciplinary and professional affiliations, contexts of use and exploitation, participant relationships, audience constraints and so on' (Bhatia, 2014: 65–66).

It is worth noting, however, that in the context of the study of genres (including media genres), the problem lies primarily in the fact that, in principle, there are no model realisations of the textual features of a given genre, only realisations that are closer or further away from a given model can be identified. This, of course, opens up a discussion, relevant, for example, in the context of research on online information genres, about how far a given textual realisation can deviate from the model, how far it can differ from it. In addition, let us also note that if we recognise the particularly important role of context and function in the analysis of genres, then even genres that appear structurally much more schematic and stabilised than, for example, news, may differ significantly in other aspects. Therefore, the starting point for the study of any genre should be structural-semiotic analysis, conducted on the basis of real texts socially considered to be representative or close to the genre (e.g. fake news identified as such by fact-checkers), from which their common, characteristic features are then abstracted to create a list of genre features and then a model (pattern).

As such, genres, as models of communicative behaviour, unconditionally possess a particular linguistic (or more broadly, sign) structure, which in turn refers us to deeper structures of meaning. The updating of a genre model (in the press, on the internet, or in a job interview) is a communicative event that pursues specific communicative goals. However, for there to be a possibility of attributing this update to a model, it must also be characterised by a specific pattern of structure, style, content, and untended audience (Swales, 2009). In the context of the ontological status of genre, it should also be noted that the discussion of genre analysed here actually concerns two planes that are often not precisely separated from each other—genre as model and genre as the realisation of that model. This use of the term is understandable insofar as the two 'dimensions' of genre condition each other: a model could not exist without communicative practices, while communicative practices could not be considered representative of a given genre if we did not have a model of it. For this reason, and because of the analytical and research traditions present in the literature, I remain here with this imprecise, though usually contextualised, understanding of the concept of genre.

In this context, it is important to remember that the study of texts with a disinformation character in the context of the genre category is also an attempt to create a model of these texts. It follows that individual realizations of news-like types, e.g. fake news, will not fully implement the features of the model, but will only approach it more or less closely. However, establishing the characteristic features of this model, modelling fake news and other disinformation messages (in the longer term) is particularly important for defining this phenomenon and determining the boundaries between malicious campaigns and other media genres, other uses of semiotic systems in the media, which is essential for conducting research in this area.

2.5 Genre in Wider Social Contexts

The modern way of thinking about genres owes a great deal primarily to the work of Mikhail Bakhtin (Bakhtin, 2010), for whom genre is a concept fundamental to the theory of communication (not only linguistic), while genre competence constitutes all cultural competence (understood as one that is symbolic in nature, i.e. connected with the ability to use symbolic systems). In Bakhtin's terms, essentially, to learn to speak is to learn to produce specific, genre-formed texts. For texts do not exist outside the category of genre, we only communicate using specific speech genres, i.e. all our utterances use concrete, relatively permanent and typical structures of a genre nature.

Human beings therefore actually communicate by means of genres and not sentences, so communicative dispositions (linguistic, semantic) are genre dispositions. Every utterance necessarily actualises a genre, and a text without a genre does not exist (which does not mean, of course, that classifying a given text into a genre is a simple and obvious matter). Genre is 'an indispensable attribute of every utterance, a more or less stabilised rule of coding a text [...], a repetitive combination of specific features' (Woźniak, 2010: 53–54). The basic function of genres, therefore, is to provide social actors with interpetable cues that allow them to understand each other's statements and shape statements that are meaningful to their interlocutors, and thus—to participate effectively in social life.

One of the basic social competences thus turns out to be generic competence, 'which not only makes it possible for members of a professional community to construct, interpret and use genres appropriately, but is also an important measure of professional expertise' (Bhatia, 2014: 164). This kind of competence, in a narrower sense, concerns the representatives of the professions (for example, journalists), whom it helps to build and maintain their discursive position as a professional group, as will be discussed further below. In a broader sense, it concerns actors in social life in general. Generic competence, as a component of discursive competence (alongside textual and social competencies), allows the identification, construction and use of specific genres in specific communicative situations for specific purposes. It regulates everyday communicative activities in such a way as to achieve specific 'community of practice' goals. In particular, generic competence makes it possible to select the appropriate type of genre according to the subject's goals; to interpret and use the appropriate generic resources so as to achieve the intended aims; to use generic knowledge to produce new texts that fit the genre model; and finally—since genres are linked to social practice—to participate not only in specific discursive procedures but also in a wide range of social practices or professional life (Bhatia, 2014).

In this way, genres are linked to discursive communites and thus to the complexes of knowledge and discourses that these communities define (Swales, 1990; Bhatia, 2013). Genres turn out to be both part and the source of a system of validation and a way of storing and keying access to specific knowledge. They are also patterns of reasoning and application of knowledge in specific circumstances, according to the norms of a community (Bazerman, 2013).

Knowledge is produced and encoded in specific genre systems, which in turn translate into specific systems of professional action. Discursive genres analysis, in the context of discursive communities, 'extends the analysis beyond the textual product to incorporate context in a broader sense to account for not only the way text is constructed, but also the way it is often interpreted, used, and exploited in specific institutional or more narrowly professional contexts to achieve specific disciplinary goals. The nature of the questions addressed in this kind of analysis may often include not only the linguistic, but also the socio-cognitive and the ethnographic' (Bhatia, 2014: 23).

Genres are thus not only highly structured and formalised communicative events of a dynamic nature, but are also embedded in a certain complex of knowledge about the world, which shapes both the form and compactness of the texts belonging to a given genre area. Genres understood in this way are produced and sustained by discursive communities with access to the aforementioned knowledge. This definition of genres leads to a very complex process of identifying and describing genre, consisting not only of an analysis of internal linguistic (semantic) structures or an analysis of the network of texts and genres that may influence the genre we are analysing. Awareness of the role of genre in the context of the production and functioning of discursive communities places it in a broadly situational context and links it to knowledge of extra-linguistic social reality. It links it to the users, especially to the discursive communities within which they function; to the values shared by these communities, to knowledge of their history and specificity; of the communicative purpose they pursue; and finally to the characteristics of the recipients of communication, whether as individuals or social groups. (Bhatia, 2013).

In the context of reconstructing fake news as a genre, this would therefore require asking a series of questions not only about its structure and relation to other genres (from which area—disinformation or real news?), but also about the contexts of occurrence of this type of texts and a detailed analysis of the group of senders as a certain community characterised by a certain knowledge of the world, their motivations and goals; and—in similar contexts—the group of recipients and the general vision of the world they are characterised by. This ambitious research agenda has been pursued, for example, in the context of the analysis of scientific texts, but let us note that in this case we are talking about a relatively closed and more easily isolated and characterised discursive community. Similar attempts have been made in the context of news, as will be discussed below, but in this case the issue proved to be much more complex, due probably to the complexity of the contexts and social groups involved in the functioning of this genre. However, there is no doubt that journalism has the status of a profession and, as such, entails the possession of a specific discursive competence, including generic competence. Does this also apply to the creators of fake news? Given the specific objectives they pursue and the significant social impact of such messages (and therefore their effectiveness), it seems that the answer to this question may be yes. From a broader perspective, it is worth looking at the discursive communities, those that produce fake news, those that share it and, finally, those that are most susceptible to its influence. Are there any shared discursive practices within them, do they refer to a shared knowledge of the world or a shared value system, do they have a communicative relationship with each other?

2.6 Discursive Approaches to Genre: Basic Assumptions and Research Frameworks

Discourse-oriented analyses of journalistic genres, especially news, incorporate basically all the currents that can be distinguished in language-oriented journalism research (Zelizer, 2004: 117–143). Indeed, questions from the fields of informal (content analysis, semiotics) and formal studies of language (sociolinguistics, critical linguistics, discourse analysis, multimodal discourse analysis) as well as pragmatic studies (narrative and storytelling, rhetoric, framing) are asked here. Moreover, within the field of discursive genre theory, the particularly important role of extra-linguistic, contextual, cognitive phenomena is also emphasised.

In research practice, however, discursive genre analyses usually begin with an analysis of textual structures. These analyses are, unsurprisingly, considered the first constitutive step, in the delineation of any genre. Let us note that such a limitation of the research field (to the text) by no means necessarily implies the abandonment of social contexts, for contemporary linguistics, as well as contemporary discourse analysis, treats language as a social phenomenon, recognising the need to study it in relation to sender, receiver, social knowledge, and function. However, and this must be emphasised, even when considering language as a social phenomenon, the starting point always remains the study of text, and genre analysis becomes 'the study of texts as social phenomena where recurrent patterns of structure and behaviour help organize structures and behaviours into comprehensible and effective forms' (Connor & Mauranen, 1999: 48) In this view, genres are initially understood as patterns realised in the semantic, grammatical, and lexical (vocabulary) features of a text at different levels of text organisation. This means that particular semantic relations or grammatical categories and relations are seen as constitutive of all genres, as well as of discourses and styles (Fairclough, 2003). The point of focus is therefore not the social situation as such, but the identifiable linguistic features of this situation, which is in turn based on a systemic-functional analysis of genre referring to semantic, syntactic and pragmatic categories. On the other hand, a text is always seen in a social context, as multifunctional, because it is simultaneously representing (producing representations of reality), setting up identities (producing social identities) and setting up relations (producing social relations) (Halliday, 1978). The study of language, although it begins with linguistic structures, is ultimately always social in nature, because it involves the question of its functions and uses in a social context, and because language is the primary means of constructing the social world.

The question of capturing language and its relation to discourse is a crucial one for discourse theory and analysis in general, especially given the complexity of the category of discourse itself and the varied ways in which this category is defined. (Bazerman, 2013; Stasiuk-Krajewska, 2018). In the context of genre research, discourse is generally understood as a communicative event that goes beyond sentence structures (Bhatia, 2013). Such an approach is inspired by a number of theories and methodologies present in the study of communication and language—from communication ethnography or ethnomethodology to speech act theory and cognitive science. The main goal of all these studies is to understand the structure and function of language used to convey meaning—either in the context of specific linguistic frameworks and formal-functional aspects of language (for example, van Dijk's analysis of news structure, Van Dijk, 1991a, 1991b, 1985) or of actual communication in an institutionalised socio-cultural context (for example, analyses of professional and academic research genres: Swales, 1990; Bhatia, 2013).

Adopting the basic assumptions of discourse theories leads to an analysis of genres in the context of different levels of discourse linked to different levels of world knowledge (as discourses by their nature are linked to world knowledge). From discourse as text (based on textual knowledge), through discourse as genre (genre knowledge) and discourse as professional practice (professional expertise), to discourse as social practice (social and pragmatic knowledge) (Bhatia, 2014). In this view, genres are one of many levels of discourse, located at a medium level of generality, yet remaining in relation to both the realisation of discourse at the most concretised textual level and the most abstract level of social knowledge.

The basic assumptions about discourse, language and their interrelationship in the context of genre theory were formulated by Norman Fairclough: 'Like many linguists, I shall use 'discourse' to refer to spoken or written language use, though I also want to extend it to include other types of semiotic activity (i.e. activity which produces meanings), such as visual images (photography, film, video, diagrams) and non-verbal communication (e.g. gestures). [...]. In referring to use of language as discourse, I am signalling a wish to investigate it in a way that is informed by the social theory insights mentioned above, as a form of social practice. Viewing language use as social practice implies, firstly, that it is a mode of action [...]. It also implies that language is a socially and historically situated mode of action, in a dialectical relationship with other facets of the social. What I mean by a dialectical relationship is that it is not only socially shaped, but also socially shaping—or socially constitutive. Critical discourse analysis explores the tension between these two sides of language use, the socially shaped and socially constitutive, rather than opting one-sidedly for one or the other. Language use—any text—is always simultaneously constitutive of (1) social identities, (2) social relations and (3) systems of knowledge and belief' (Fairclough, 1995: 54).

The above excerpt highlights the most salient assumptions of discourse theory that underpinned discourse-oriented genres theory and analysis. Discourse, identified with the use not only of language but also of other sign systems (which became the basis for the development of multimodal discourse analysis and the formulation of the notion of multimodal genre), is understood as social practice, as action. Thus, unlike in the theoretical approaches of Michel Foucualt or Erensto Laclau and Shantal Mouffe (Foucault, 1972; Laclau & Mouffe, 1985), social practice is not only a product of discourse, an area of disclosure of its rules, but is also a place of the production and evolution of the discursive. Discourse is not a set of rules, but a communicative and social action. Fairclough is well aware of these two fundamentally different approaches to defining and analysing discourse ('One is predominant in language studies: discourse as social action and interaction, people interacting together in real social situations. The other is predominant in post-structuralist social theory [...]: a discourse as a social construction of reality, a form of knowledge', Fairclough,

1995: 18), yet at the same time he emphasises that the understanding of discourse he proposes, in the context of genres research, goes beyond this opposition, because it 'subsumes both of these, and indeed sets out to bring them together. The first sense is most closely associated with the interpersonal function of language, and with the concept of genre [...]. The second sense is most closely associated with the ideational function of language, and with discourses—notice that in addition to being used as an abstract noun for this general view of language in social use, discourse is used as a countable noun (a discourse, several discourses) as a category (alongside 'genre') within the intertextual analysis of texts' (Fairclough, 1995: 18). And although the issue of the fundamental divergence between the two schools in the study of discourse is unlikely to be resolved in such a simple way, the above statement presents a very important insight that makes the aforementioned division somewhat less clear-cut in analytical practice. On the one hand, social practice of communication (text) is what we can observe and study as discourse, on the other hand—every use of language is the product of socially constructed and socially negotiated processes of meaning construction. In this sense, communication is always in relation to the social, to culture and ideology, and is therefore discursive. One should not study language (or any other communicative system) merely as a system that is a structure of certain elements and rules that regulate the coexistence of these elements, but should look at communicative systems as semantic systems, thus generating certain meanings, certain interpretations of reality. And semantic systems, as already implied by the definition of the sign as such, are necessarily discursive. In summary, discourses are ways of representing the world which can be identified and differentiated at different levels of abstraction. Texts articulate different discourses together, in various ways. Discourses can be differentiated in terms of semantic relations (synonymy, hyponymy, antonymy) between words as well as collocations, assumptions, and various grammatical features (Fairclough, 2003).

Understood in this way, the language-discourse relationship allows elements of Critical Discourse Analysis (CDA) to be introduced into the analysis, whose primary research endeavour is to analyse the use of language as discourse, with the aim of revealing the role of discourse in the production and maintenance of the social world, including relationships based on unequal power relations and thus leading to discrimination. In order for an analysis of language as discourse (or discourse as language) to be possible, several fundamental assumptions must be made about language (Richardson, 2009: 10–14). If, as has been mentioned many times before, language is a social phenomenon, it is also a tool for the production and maintenance of social identity—both individual and (more importantly in the context of genres) group, community or professional identity (recall the specific role of genres in this process described earlier). In producing social identity, language is at the same time associated with power—in the narrower sense (statements of certain social groups have more power, are better heard and more often listened to than those of representatives of other social groups) and in the broader sense (establishing dominant ways of talking about the world). Since language is related to the distribution of power, any use of language is also political—language is not a neutral description of reality, but, as a medium of power, establishes relations of power and domination. All these functions of language could not be realised if language were not always active, in the sense that the use of language involves doing something, has a pragmatic aspect, realises a certain function (for example, journalists, using the news genre, report on the state of the world). In this context, each genre, realised as a linguistic practice (news, but also fake news) produces a certain social relation and the identity of individuals and groups involved in this relation; it brings a certain real pragmatic effect. It also actualises and creates power relations, understood as social domination (e.g. of an expert group of journalists or the author(s) of fake news, as a source of unavailable knowledge about the world) or symbolic domination (e.g. by actualising, for example, the so-called news ideology or, on the contrary, conspiracy theories as discourses explaining the world). Linguistic (communicative) practice is therefore always a discursive practice, since the existence of language and discourse mutually condition and enable each other.

2.7 News as a Genre

Although, as van Dijk writes: 'It is striking, to say the least, that we know much more about poetry, drama, novels, myths, or storytelling, for instance, than about the news reports we read in our papers or see on television every day' (Van Dijk, 1989: 121), it has to be admitted that news is a relatively well-described genre in the literature. This is probably due, on the one hand, to its relatively far-reaching schematisation, and, on the other hand, to the special role it plays both in the construction of the journalistic profession's self-description and the identity of democratic societies more broadly. Analysing this phenomenon from the point of view of theoretically oriented discourse analysis, it is worth noting that the basis of the semantic chain constituting news as a genre is the category of truth, which, as can be shown, constitutes one of the essential nodal points of the discourse of modernity (as can also be seen in the special role attributed to scientific cognition as the method that has access to true knowledge of the world) (Stasiuk-Krajewska, 2018). Journalists take over the role previously performed by priests or intellectuals (Fiske & Hartley, 2004), as those showing the truth about the world build their position in relation to scientific discourse.

It is important to remember that the reference to truth as a constitutive value of news is of interest here not as a factual state of affairs, but as a discursive procedure. It is not a question of whether news stories actually represent the truth about reality (which, ultimately, in a deeper philosophical sense, is probably impossible), but of the fact that they present themselves as representing this truth about reality and are (at least in an ideal communicative situation) interpreted as such by their audiences. This is an important reservation because let us note that in this sense fake news is also constructed and treated by its recipients as presenting the truth about reality, referring to facts that are intentionally concealed. Truth, in this view, is not a relation to reality, but a specific functional textural and discursive structure. This observation seems to confirm the thesis that fake news is similar to journalistic news in terms of genre (where genre is understood here broadly, not only as a certain linguistic structure, but also as certain semantics and thus discourse).

In discursively oriented research on news,[1] it is primarily assumed (in line with the general assumptions of genre research outlined earlier) that two main determinants of news should be taken into account: the semantic system that encodes it and the social forces that determine how news is encoded on the one hand and decoded on the other (Hartley, 2013; Miller, 1984).

[1] The literature sometimes distinguishes, on the basis of subject matter, between hard news and soft news (Bell, 1991: 14). Hard news deals with serious accidents, crimes, conflicts, violence, terrorism or natural disasters, and therefore issues constructed as important and serious, significantly disrupting the social order and sense of security. Soft news, on the other hand, has a gossipy, causal and, in this sense, tabloid character. While the choice of subject matter is certainly important, the distinction does not seem to be fundamental to the structure of the news as a genre, especially since the fake news genre includes both 'serious' and 'entertainment' material (although the latter is in the minority).

Regarding structural issues, both formal (grammar, syntax or lexis) and less formal (storytelling, frames, textual patterns and narrative) aspects of language are analysed. Of particular interest is the structure of news stories—the inverted pyramid is most commonly referred to here, but also the Kabob, Martini Glass or Narrative structure (Dai et al., 2018). The individual elements of the structure perform specific functions (e.g. introducing the main themes, attracting the viewer's attention or presenting the context of the event). What is particularly interesting is that most news stories represent an inverted chronology (Bell, 1994), placing what is most important or potentially most attractive to the viewer first, rather than what happened first. The clearest illustration of this principle is the title or headline, a structural element without which it is difficult to imagine a news item. Often, the functions of other structural text elements are also determined precisely in relation to the title. They may, in particular, provide a more detailed description of what is contained in the title (elaboration and reiteration); explain the causes or justify the non-standard events signalled in the title (causes); describe the consequences of the events that have occurred (consequences), provide a broader context for the events presented (contextualisation), or evaluate these events, by introducing another voice rather than through a judgement made by the journalist (attitudinal assessment) (White, 1998).

In addition to structural elements in the context of the news genre, there is also discussion about specific styles that can be observed in news stories, including in particular narrative and expository styles. 'A narrative section in a story usually describes surroundings, characters, and a sequence of events in a chronological order […], so that the reader can easily visualise the story with great detail. An expository section is meant to provide information in a concise manner and usually answers the so-called '5W1H' questions: what are the events, who are involved, where / when / why / how did the events happen' (Dai et al., 2018: 61). The two styles, although functionally distinct, nevertheless share some common features, namely they are considered objective, detached and factual (Bhatia, 2014: 93).

Styles are also sometimes referred to as voices. The category of voice refers to a particular pattern of language use and the presence (or not) of semantics that are associated with evaluation. The classic voice present in news (especially in hard news) is referred to as 'reporter voice'. It should be 'factual', 'neutral' and free of 'subjectivity'—so it is characterised by a significant restriction on the author's use of explicity attitudinal menaces, especially in terms of expressing judgement (Martin & White, 2005).

Note that using such a voice, and thus writing in the style characterised above, is considered an essential professional competence and is often part of journalistic education.

There is also no doubt that the news story is a specific narrative model. Typical news narratives usually include (White, 1997):

(a) elaboration, a single sentence or group of sentences provides a more detailed description or exemplification of the information presented in the headline/lead;
(b) cause-and-effect, one or more sentences describe the causes, reasons, consequences or purpose of the crisis presented in the headline;
(c) justification, one or more sentences present evidence or justification that supports the claim made in the headline;
(d) contextualisation, one or more sentences place the events of the headline in a temporal, spatial or social context
(e) apprisal, the elements of the headline are evaluated, usually by some external expert source, in terms of their meaning, emotional impact or by reference to some value system.

The relationship between genre and narrative is not clear (reading news as narrative), but certainly both categories refer, on the one hand, to the use of a language geared towards the realisation of a specific goal and, on the other, to actualisation through this use of specific interpretations of the social world (Fairclough, 1995).

However, as has been mentioned many times, discursive analysis of news is not only an analysis of structures, narratives, or stylistics, but also of the ways of interpreting the social world actualised by the genre. What ensures the relative communicative stability of the genre is not only the internal structures of the texts, but also the social system, and this includes the institutional system of which the genre, as a pattern and as realisations of the pattern, is a part.

Just as the model of the scientific article could not exist without the constitution of science as a distinct field and the emergence of procedures, institutions, etc. associated with this field (Swales, 1990), neither could the news model exist without the distinct field of professional journalism. The journalistic practice and media theory of news is intertwined from the outset with notions of civil society and democracy (Schudson, 1978). This narrative is based on two main assumptions: firstly, that a properly functioning democracy requires informed citizens, and secondly, that it is the news that provides these citizens with the information they need to fulfil their civic duties, and thus for the proper functioning of democracy. It is out of this narrative, out of this construction of journalism and its social role, that a set of norms for journalistic ethics, including the crown genre of journalism, which is news, is derived. These norms are then, in line with the basic tenets of discourse theory, translated into structural and stylistic features of texts that fall into the news genre.

The values structurally encoded by informative and rational news, relating to truth understood as conformity to facts and objective (scientific) knowledge of the world, are at the same time the core of journalistic professional culture, and produce certain practices and beliefs, primarily related to objectivity and the assumption that news should be professional, unbiased, honest, and reliable (Deuze, 2005). These beliefs then become an important element in the self-description of the professional discursive community of journalists, as reflected, for example, in codes of professional ethics (Stasiuk-Krajewska, 2018).

Thus, one important function of news is to build and maintain the social position of a professional community of journalists. However, this is not the only function. In light of the assumptions of theoretically oriented discourse analysis as well as, to some extent CDA, news produces and sustains a certain interpretation of reality that constitutes the basic semantics of modernism (Stasiuk-Krajewska, 2018; Hartley, 2013). The news thus conscripts a certain type of generally acceptable social knowledge, and constitutes cultural values that are perceived as certain and socially relevant. This knowledge and these values are productive in the sense that they then produce a kind of horizon of expectations, a range of acceptance for subsequent discursive procedures and practices. 'Whether through newspapers or through television, news discourse influences the fundamental contents and principles of our knowledge and (other) social representations. True, we may not believe everything we read in the paper, nor do we necessarily feel influenced by the reported speech of our president, but the news nevertheless will influence what we actually think about (even critically), what we find relevant or irrelevant, which people are found prominent or marginal, and which events are deemed interesting or uninteresting. In sum, news discourse has an impact on social knowledge, attitudes, and ideologies, despite the social or political differences of the readers. If it does not always directly influence our opinions, it may partly determine the principles and strategies of our social information processing, that is the interpretation frameworks we apply in the understanding of social and political events' (van Dijk, 1989: 134–135).

The category of journalistic ideology is also used in the context described above. This strand of research and analysis developed within the constructivist view of journalism (see Chap. 6 of this book) and focused primarily on news values, which will be discussed later in the book. From the point of view of the present discussion, particularly relevant are those works that dealt with ideology in news reporting by identifying and describing certain linguistic or discursive elements. In doing so, it was assumed that discourse was a field of both ideological and linguistic processes, and therefore linguistic theory could be used to uncover ideological processes in journalistic work (Trew, 1979a, 1979b). It has also been argued that the media (the printed press) present a particular vision of the social world, a particular ideology, which we can study from the perspective of the broadcaster (e.g. by interviewing media representatives or by analysing the guidelines on which a newspaper's editorial policy is based); by analysing the process of transmutation, the transition from the world of events to the publication of a newspaper article; and finally, by analysing the media product, i.e. the texts themselves (Hodge, 1979). The specific ideology of the European and American media is also evident in comparative contexts—studies that show the differences in the structures and functions of news messages, and thus the vision of the world updated by and through them, in other cultures (Kopytowska, 2015).

News is, as has been discussed, the result of a complicated set of journalistic practices, the first step of which is the selection of events and topics according to socially constructed categories. These categories are part of the discursive field of modernity, the dominant contemporary way in which the social world is constructed. They update, for example, the discursively constructed distinction between truth and falsehood, but also, being discursive in nature, the boundaries between information and entertainment or information and advertising; citizens and consumers, etc. These boundaries, which form the basis of the discourse of contemporary democratic societies (the dominant worldview), are treated as unambiguous and universally binding—they are part of hegemonic discourses (Laclau & Mouffe, 1985) that represent symbolic power. By inscribing itself in these hegemonic discourses, journalism simultaneously builds its strong position in the field of modern societies, legitimises itself and differentiates itself from other professional communication practices (related to the market or entertainment, for example). To reiterate—a special role in these processes falls to news as a genre. 'One way of seeing news is as a form of social regulation, even a form of violence: news reduces a complex series of events whose relationship may not be terribly clear to stories, imposing narrative order upon them. And it is not simply the relationship between an actual series of events in a particular order, and the story about them. [...] Making news is a heavily interpretative and constructive process, not simply a report of `the facts. This does not mean that news narratives are just the same as fictional narratives: news narratives [...] have a 'referential intention' which makes them open to questions about the relationship between story and actual events, questions of truth. They also have, one might say, an 'explanatory intention' which we can liken to 'focalization': to make sense of events by drawing them into a relation which incorporates a particular point of view' (Fairclough, 2003: 84–85).

At the communicative level, such assumptions about self-evident 'dawning truths' are realised through presuppositions, which are a kind of implied meaning embedded in the text. At the basic level of language research, the presupposition of a sentence is understood as a condition that must be fulfilled in the world for a sentence to be considered meaningful. In the context of news, however, the category of discursive presupposition is more relevant, i.e. an element of knowledge about the world that is taken for granted, given within a particular discourse. In this sense, the assumptions about the world discussed above can be considered presuppositions of journalistic discourse. Their manifestation can be traced in the structure and semantics of news (Bekalu, 2006).

At the most general level, therefore, it can be said that news legitimises the dominant vision of the world. This can also be clearly seen in the context of the typical topics that are associated with the model realisation of the news genre—with hard news. They concern what is constructed as socially relevant (note—this is in fact the basis of the distinction between hard and soft news—the latter concerns issues considered trivial, socially irrelevant), but also what disturbs the expected, 'natural' order of things, a vision of the world in line with the dominant interpretation of it. 'The generic structure of the 'hard news' requires that the reporter construes events and statements in terms of the purported risk they pose for the social order and that this construal is an act of ideologically and culturally determined interpretation. [Narrative impulses] act to categorise events and issues as more or less disruptive, transformative, transgressive or destructive of the social order. But there is more to the 'hard news' story than this 'narrative' of social-order disruption. The structure of news reports of this type provides a powerful rhetorical device for representing these ideologically determined choices about status-quo disruption as natural, necessary, and value-free. The structure operates on multiple levels to achieve this outcome' (White, 1997: 23). Aberrant damage, adversative rearrangements of power relations and normative breach, the main thematic categories of hard news (White, 1997), undermine trust in the causality and control of the world, the rationality of power mechanisms in democratic societies or the stability of moral rules that govern the world. These semantics (causality, control, the rationality of power, the relevance of moral rules, etc.) are at the same time the underlying semantics (preconceptions) on which modern societies are built. By presenting events that challenge this interpretation of the world as news (significant, but at the same time overwhelmingly negative), the media simultaneously reinforce the hegemonic discourse—creating the impression that everything that goes beyond the rules of this discourse is abberative, disturbing, and even dangerous. This mechanism perfectly explains the well-known negativity of the media, and allows us to see in a different light the paradoxical feature of tabloids, which—while presenting scandalous content—remain ideologically conservative. For these, scandalous contents actually reinforce the conviction of the relevance and stability of the dominant rules of social life. The discourse of celebrities, presenting their excesses but at the same time subjecting them to moral evaluation, is one of the more conservative media discourses.

In this vision of the social world, legitimised by the hegemonic discourse, which is (re)constructed by the news (2), society is furthermore perceived as fragmented, divided into separate spheres such as sport, politics or education (note, as an aside, how much negative coverage in social media or online forums was aroused by the inclusion of social issues in reports of the 2024 Paris Olympics); composed of individuals who control, or at least should control, their lives and the world around them (a clear vestige of modernist individualism); hierarchical by its nature, with people or groups who are justifiably more empowered than others; and finally, consensual, one in which agreement, or at least rational conversation, should be sought (here we can also see a characteristic concept of modernism—the vision of a rational public sphere).

The above features and functions of news are also analysed in the context of narrative theory. If news is a narrative, then, as such, it transmits certain cultural or ideological meanings, it naturalises a certain interpretation of the world. This is a feature of narrative as such, for the concept of 'narrative' refers to how discourses (re)construct social reality. Narratives are, after all, considered as basic elements of discourse or even identified with discourses (see, for example: Propp, 1968; Greimas, 1971). Just as established narrative patterns work in popular culture, so in professional journalism the use of typical journalistic narratives builds the position of the sender, but also a kind of consensus with the recipient, who feels cognitively safe in a familiar and recognisable narrative context.

In the context of the reflections formulated above, it is essential to emphasise the fact that they do not in any way lead to a denial of the relevance of news as a media genre for audiences or society more broadly, nor do they have the character of a conspiracy theory of media and journalism. 'First, the impartiality, objectivity, neutrality and balance which form the bedrock of editorial ideology are no sham. They are required if news is to act alongside the other agencies in naturalizing dominant ideology and winning consent for hegemony. Next, the mutual confirmation of ideological meanings between news and agencies like the family and the school is no deliberate conspiracy to 'dupe' the public. It is the product of a complex historical process and is deeply embedded in the discourses through which we learn to interact with the world, and in which we make sense of it. This goes just as much for news-people as for anyone else. It suggests, further, that news is not a producer of ideological meanings in the sense that they are originated in it and nowhere else. News re-produces dominant ideological discourses in its special areas of competence. Finally, the ideology of the news is not a 'partisan' ideology. On the contrary, the purpose of news ideology is to translate and generalise, not to choose this opinion or that. In other words, news naturalizes the (fairly narrow) terrain on which different sectional ideologies can contend—it constantly maps the limits of controversy' (Hartley, 2013: 61–62).

When asking about fake news in this context, one should first ask whether texts of this kind appeal to a different ideological, discursive framework (in the broad sense of the word) than real news texts? Does the knowledge they (re)construct have a fundamentally different character? On the one hand, conspiracy theories are the opposite of scientific knowledge, which they undermine and depreciate. On the other hand, however, they map onto the mechanisms of scientific thinking and cognition, and construct themselves as presenting true knowledge about the world. In the case of fake news, both the subject matter (which refers to what is also of interest in 'hard news') and the structure (reminiscent of the news genre) also seem to refer to similar values: control of the world through knowledge of it or a kind of stability guaranteed by universally applicable moral rules. Of course, the authorities, the sources of knowledge, and the negative and positive protagonists of these narratives are different, but the (re)constructed structure of the discourse, the interpretation of the social world seems disturbingly convergent.

2.8 News Discourse

The most elaborate, structured and inspiring (for example: Pan & Kosicki, 1993; Yarlott et al., 2018) analysis of news as a genre representing discursive features mediated by language was proposed by Theun van Dijk, already cited. In his model, van Dijk applied tools not only from linguistics, semiotics or speech act analysis, but also derived from anthropology, sociology or psychology, building an analytical system that also gives a theoretical and methodological basis to CDA (mainly: van Dijk, 1985, 1987, 1988). The critical aspect is mainly evident in the fact that van Dijk's analyses focus on the ways in which journalistic texts support the unequal distribution of symbolic power in society. Of particular relevance to the present analyses is the assumption that discrimination (gender, ethnicity, class, etc.) can be found in language, including the language of news.

Van Dijk's conception presents a relatively precisely described analytical scheme, which can then be applied to the analysis of news, in direct relation to theoretical assumptions about discourse. These assumptions primarily conceptualise discourse in terms of three dimensions or perspectives (Fairclough, 1995): text, discourse practice (encompassing cognitive issues related to the processing of discourse in the processes of its production and consumption), and socio-cultural practice (concerned with context in the broadest sense, including, for example, institutional or professional knowledge). In other words, in analysing a text one examines, firstly, how its structure and the discourse strategies employed in it are used to emphasise a particular theme. Secondly, the cognitive aspect is taken into account, which includes the processes of text production and reception, as well as the individual cognitive processes involved in actualising discourses. The social context, finally, refers to the discourse construction that develops in society in relation to a given topic.

In particular, the important shift from text analysis to discourse analysis is evident in this premise, as well as the multifaceted treatment of discourse and the attempt to answer the question of how social relations and processes (e.g. the reproduction of racism) are realised at the micro level through routine practices, including textual practices. The general framework of discourse analysis translates into a framework for analysing news discourse in terms of the structures of news texts, the processes of news production, and their understanding by broadcasters and audiences. The theoretical aim of the analysis is to show the relationship between texts, processes of production, and processes of understanding, as well as between these and the wider social practices in which they are embedded.

The first step of text analysis is grammatical analysis, which is concerned with the abstract levels of discourse: the systems and rules for combining signs, their meanings, and therefore their functions. It is also important to note that grammatical analyses do not refer only to the sentence level, but primarily to a higher level—the over sentence sequences of the text. This level of analysis deals with phenomena that van Dijk (1985, 1986) refers to as local structures (microstructures) and includes both sentence analysis (not only morphology and syntax but also semantics and lexicon) and sequential structure analysis (relative syntax: cohesion analysis and relative semantics: coherence analysis). In practice, analysis at this level means studying the semantics of words or the discursive consequences of using certain grammatical structures (e.g. the passive voice or question sentences). Microanalysis thus identifies the syntactic and lexical features of a text's style and the rhetorical features of news, for example those that give it the specificity of informationality or objectivity. At the semantic level, it refers to the preconceptions encoded by the text; at the syntactic level, to the choice of sentences and the relations between them (coherence); at the stylistic level, to the modality of words; and finally, at the rhetorical level, to the metaphors used.

The second level of text analysis is the level of global structures, including macrostructure. Macrostructure, a concept very relevant to van Dijk's general model of discourse analysis (van Dijk, 2019), is semantic in nature and refers to the overall content of a text—its 'thematic' structure. The macro-structure of a text is its overall organisation in terms of a theme or themes. It is a hierarchical organisation, in the sense that we can identify the dominant theme of the whole text (and summarise it as a single theme), and then describe it in terms of several less general themes, which in turn can be broken down into categories of even more specific themes, and so on. Macrostructures can also be understood as certain ways of structuring content that are typical of the news, however, not formal, but relating to semantics. In this sense, they would be certain kinds of narratives that constitute the personality of the sender or the interpretation of the world present in the news. These narratives, as certain generalised ways of interpreting the world, actualise those elements of discourse that are constitutive of the news as a genre. In this sense, the semantic macrostructures would be in correlation with what it defines as *news values.* At the same time, macrostructures, at the level of function, allow the recipient to locate the text in an adequate communicative situation, in this case the communication of news. It can therefore be assumed that, at this most general level, thematic structures would be a certain semantic way of organising content, such as generalisation or reconstruction, which enable the recipients to orientate themselves relatively easily to the meaning of the whole text. This organisation is formally expressed, for example, in the form of a headline or a lede and their function in the context of the semantics and the way the whole body of information is received. The schematic structure also organises the thematic content, thereby giving it a certain rank. In the case of news, the principle of relevance is dominant, which translates into an overall inverted pyramid structure.

The second structure of a global nature is the superstructure. It, in turn, is formal in nature, a schema for the organisation of content. The schematic structure of a given text type is defined in terms of the ordered parts of which it is constructed. In van Dijk's approach, a news story consists of a headline, the lead; then an 'events' element, which covers the main events in the story; and an element that presents verbal reactions to the story or a commentary element (the latter two are not always applicable). Each superstructure element corresponds to a more general theme in the macrostructure (for example, the news headline formulates the overall theme of the text). This category points explicitly to the assumption that news stories also have a conventional form—a schema that organises their overall content—so superstructures, although operating at a local level, can be described in terms of conventional linguistic rules or strategies (much like genre as a pattern of speech, van Dijk, 1985).

In analytical practice, macrostructures and superstructures are inextricably linked, confirming one of the characteristics of discourse as such, which actualises itself through the entanglement of meaning and structure. Of course, such a connection makes the operationalisation of van Dijk's concept much more difficult, as the two levels overlap and condition each other.

The macro-structures and micro-structures present in the text are intrinsically linked to mental models of events and situations, and thus to the cognitive area, which plays a very important role in van Dijk's conception. Cognitive processes (van Dijk, 1989) are based on organised social knowledge, both knowledge of language, discourse and communication, and overall knowledge of 'the world'. They involve scripts, models or theses of strategic processing and control, social representations, as complex as stereotypes or ideologies. Van Dijk attaches particular importance in this context to the category of model. Models are what provide a kind of 'link' between the semantics of the text and the cognitive processes taking place on the sender and receiver side. 'Language users are not primarily aiming at the construction of such a semantic representation of the text (possibly with some information about its surface structure style, when relevant), but rather trying to understand what the text 'is about', for instance an action or an event. It is the representation of the situation the text 'refers to' which language users represent in a mental model, a so-called 'situation model'. Models, thus, are what people imagine to be 'the case' when they understand a text or participate in a conversation. In this way, we obtain so to speak a double representation of meaning in memory, viz., a meaning of the text, on the one hand, and a meaning of the event the text is about, a fragment of the world, on the other hand' (van Dijk, 1989: 125). Models are the cognitive schemas in which we place the information we receive. These schemas are a hierarchical network of categories correlated with the semantic structures of the text: a Setting (Place and Time), Circumstances, Participants with different roles (Agents, Patients, Objects, etc.), Events or Actions, and their respective evaluative modifiers (Good, Bad, Pleasant, etc.) (van Dijk, 1989: 126). Models are both the schemas that journalists use in interpreting events and source texts and then try to convey to their audiences through the way they write news; and the schemas that audiences (readers) use in interpreting news. The cognitive perspective sheds light on discursive practices and their relation to the texts.

The final stage of the analysis, the analysis of broad social contexts, deals with issues that have already been addressed earlier in this chapter—knowledge of the world, both generalised and understood as a constitutive element of discursive communities.

The analyses of fake news carried out so far in the news discourse paradigm indicate the cognitive effectiveness of this model. The realisation of this model is the study, which was conducted on 435 fake news stories in Polish, which were labelled as 'false' and therefore considered to be fake news, by AFP in Poland between January 2020 and December 2022

(Stasiuk-Krajewska, 2024). This study, which serves as a kind of reference point for the analyses presented in the following two chapters, dealt with the characteristic structural and semiotic features of fake news as a genre. (in line with: van Dijk, 1983).

In terms of structural features, the characteristics of the sender (intentions, goals, mood), the relationship with the receiver (confidentiality, proximity, power) and the type of social situation were taken into account; while in terms of semiotic features, the theme, narrative (story) and characters were considered. The analysis was based on the study of both textual and visual elements of fake news. In terms of language, consideration was given, among other things, to the structure of the headline (as a particularly important formal element), the vocabulary encoding emotions and vocabulary and elements suggesting the knowledge and objectivity of the sender (for example, numbers) or the presence of emoticons. In terms of visual analysis, attention was paid, among other things, to the presentation of the protagonists (in terms of their gender or represented profession) and the spatial and factual contexts of their presentation. Formal elements of the visual messages, such as framing or colour, were also addressed.

In the first area, the structural one, the way in which the receiver constructs himself and his relationship with the sender was analysed, as well as the way in which the communicative situation in which a given communicative practice is actualised is constructed. The analyses showed that the sender constructs himself as objective and emotionless, although this is not an explicit construction. There is also a clear tendency to reveal judgements (especially through an ironic attitude to things, wistfulness). It is not possible to state unequivocally that emotions are not encoded at all—in about 1/3 of the messages (depending on the analysed variable) these emotions are present (they are mainly negative in nature). However, the study indicated a relatively low level of emotionality of fake news in relation to the common perception of the subject. In the relationship with the recipient, the sender constructs himself as an authority with access to exclusive knowledge. Also, however, willing to share this knowledge and thus create a community. As can be seen, the construction of the broadcaster and its relationship with the audience cannot be considered identical to the ideal construction (role) of the news journalist. On the other hand, many structural elements (e.g. the title, the reporter's photo, etc.) directly relate to the situation of receiving a news text.

When it comes to the area of semantics, fake news themes, like news themes, are topical and relevant to the audience. This is not surprising, as fake news 'reacts' most often to dominant media topics, or more broadly, issues that move public opinion. The issue of narrative is characteristic—the updated narratives explicitly refer to conspiracy theories, building a coherent vision of the world. Representations of people play an important role—this is certainly a marker of personalisation on the one hand, and on the other a clear production of the opposition between us ('ordinary people') and them (politicians, functionaries, etc.).

At the same time, the study showed that fake news, like news, supplements the text message with a visual message, which mostly imitates a daily photograph (86%). The majority of such messages also have a title (77%) in the form of an announcement (76%), which performs functions analogous to those attributed to it in real news.

2.9 Discursive Context of 'digital news'

Van Dijk's conception of news genre, presented above and operationalised, can be supplemented with some elements drawn from the theory of another significant discourse researcher also located in the CDA stream, Norman Fairclough. In Fairclough's theory, as in van Dijk's, the study of discourse must be mediated by the study of social practices, understood as relatively stabilised forms of social activity: 'social relations, social identities, cultural values and consciousness, amongst others, are in part discoursal, but that does not mean that we theorize and research social relations, for example, in the same way that we theorize and research language. They have distinct properties, and researching them gives rise to distinct disciplines' (Fairlough, 2001: 234) Starting from these assumptions, Fairclough analysed certain types of journalistic messages—such as interviews or news reports—in terms of their dependence on texts from other contexts, such as government reports or press releases, focusing on intertextual aspects, and thus on the interrelationships between texts with different characteristics (Fairclough, 1992, 1995)

However, genre as such was also the object of his interest. Starting from CDA's characteristic assumption that discourse should be analysed at three levels: text, discursive practices, and sociocultural practices, Firclough sees genre as one type of activity that constitutes the social through the symbolic (alongside discourse, referring to how a text creates an interpretation of the world, and style, which in turn indicates how the author/

speaker constructs his or her identity (Fairclough, 2000). 'The discourse/semiotic aspect of a social order is what we can call an `order of discourse.' It is the way in which diverse genres and discourses and styles are networked together. An order of discourse is a social structuring of semiotic difference—a particular social ordering of relationships amongst different ways of making meaning, i.e. different discourses and genres and styles' (Fairclough, 2003: 207). The distinctive feature of genre in this context is its internal formalised structure, which forms the basis for its distinction. In the case of news reports, this is: headline + main paragraph + 'satellite' paragraphs (Fairclough, 2003). The analysis of the internal relations of texts concerns semantic, grammatical, dictionary, and phonological relations.

However, the overall perspective of genre studies is much broader, encompassing 'Activity, Social Relations, and Communication Technology—what are people doing, what are the social relations between them, and what communication technology (if any) does their activity depend on?' (Fairclough, 2003: 71). From the point of view of the study of media genres, the category of communication technology, hitherto not strongly exposed, seems particularly interesting.

At a basic level, this involves the classic distinction in communication theory between two-way and one-way communication and mediated and unmediated communication. On this basis, it is possible to generate a classification of communication situations that takes both planes into account; some of these situations refer to communication mediated by (mass)media of various types. However, another observation, of a more general nature, is more relevant: One way in which genres differ from one another is in the communication technologies they are specialised for, and one factor in changing genres is developments in communication technologies: the development of new communication technologies goes along with the development of new genres (Fairclough, 2003). One example, of course, is the development/modification of media genres faced on the Internet. The expansion of various from multimodality, i.e. the combination of different semiotic modalities, especially visual (Kress & Van Leeuwen, 2001), and non-sequentiality, in that the format offers a range of choices to take a number of different decoding paths, are considered to be the most characteristic features of this type of communication.

The question of online genres, in particular the question of the modifications to which online news genres have been and are being subjected (Wahl-Jorgensen & Hanitzsch, 2020), is of course relevant to the analysis of fake news as a genre. Digital genre (and digital news) are of great interest to researchers (Allan, 2006; Eriksen & Ihlström, 2000; Xia, 2020; Belcher, 2023; Schäfer, 2023). In this context, a number of important phenomena are discussed, which relate primarily to the formal aspects of texts. However, it remains an open question whether such formal changes translate into deeper structural and, more importantly, functional changes. If we assume that the basis for the functioning of a news genre is a communicative contract related to the intention of the sender and the expectations of the receiver, which underpins news structures and semantics, we may find ourselves in two situations in the context of online genres. Firstly, if changes in the genre model do not undermine the basic function of news (in terms of the sender-receiver relationship), so if the receiver still reads a given message as news (and, as the result, has certain expectations), then it is difficult to speak of a genre change. If, on the other hand, there is a change in the communicative contract between sender and receiver, a change in the function of the text and its resulting structure, then we would have to recognise the emergence of a different genre, which should be analysed separately from the news genre. This should then also be reflected in the nomenclature and in research practice—for if we still use the news category, it means that we still stay in the context of traditional journalism practice, even if we see some (more or less significant) modifications to it.

One of these modifications is considered to be the fact that news on the internet is layered, Aitchison & Lewis, 2003) and in a permanent state of flux (Eriksen & Ihlström, 2000). It is related to a certain variability in the structure of the text. Firstly, a news item on the Internet consists of several layers, which can be manipulated by the recipient at will; secondly, its structure can also be changed linearly, already after publication, through interventions by the author or the editors. In connection with the evolution of news described in this way, terms such as annotative journalism (Paul, 1995) or contextualised Journalism (Pavlik, 2001), and especially 'ambient Journalism' ('a system that offers diverse means to collect, communicate, share and display news and information, serving diverse purposes', Hermida, 2010: 301) already several decades appear. In traditional news it was the journalist or editor who made the unalterable choice of the degree of detail and the breadth of data presented, in this case it can be done by the reader. Undoubtedly, this structure is less stable, weakens the boundaries between the texts, and puts more power in the hands of the reader. Let us note, however, that still, at the level of the text, the reader is given a certain structure, beyond which he or she may or may not go, and any changes made by the editors take place within that structure. In view of the above, the question of social practice turns out to be fundamental. Firstly, how often does the reader take the opportunity to expand the context, and secondly, what effect does this have on the reader's interpretation of the primary content? How far are editorial interventions into the text after publication and to what extent are they structural?

It would therefore also be appropriate to ask how discursive practices are realised. There are interesting observations about the shifting of news from the realm of fixed representations towards 'a kind of constantly utated flow of 'news now' (Sheller, 2015: 13). The democratisation of media use, the increasing role of the reader in the function of the broadcaster, is also often noted (e.g. Jenkins & Deuze, 2008), but let us note that the content provided by readers to the media is efficiently placed within the framework and communicative structures of these media.

The analysis of the possible news genre on the Internet also requires reference to a third level of discourse analysis—socio-cultural practices. Of course, with social change and technological developments, the temporal and spatial dimensions of news broadcasting and delivery change (e.g. in the context of Mobile Journalism), the forms of presentation change (e.g. from press format to television format), and finally the ways of perceiving the world change. These latter changes, however, are not so important for news if, according to the assumptions made here, we consider that news structures refer back to the basic and therefore stable semantics constituting the vision of the world in modern societies. To put it another way, as long as we speak of news or Journalism, no matter what adjective we attach to these terms, we still remain within the traditional semantics of modernism concerning the functions of the media, and therefore also the functions of journalism and news. We remain within the same discourse type, and thus inside 'relatively stabilised configurations of genres and discourses within the order of discourse' (Fairclough, 1995: 66). The above observations are confirmed by communication practice—research shows that the content of media or online news channels (such as BuzzFeed, for example) tends to resemble traditional news structures (Wu, 2016; Tandoc & Jenkins, 2017; Hurcombe et al., 2021).

As it seems, then, online journalism, defined by 'emerging features of multimodality, hyertextuality and interactivity' (Song & Lin, 2012) has not led to profound changes in the structures and functions of news. In terms of genre research, there are certainly new challenges (related, for example, to the aforementioned structural instability or the difficulty concerning the description of media uses or the dynamic context of message perception, and new terms or taxonomies (Xia, 2020), but it is difficult to speak of the constitution of new genres, especially in the news function.

An interesting exception is social news, treated, it seems legitimately, as a new genre. Social news remains in a structured relationship with the logic of social media, both in terms of communicative conventions and the construction of the relationship between sender and receiver (Gibbs et al., 2014; Burgess, 2006). In the context of social news, there is even discussion about the constitution of a new news value of shareability. Let us note, however, given the framing of news values in the context of traditional news, that shareability would be the result of the effective implementation of traditional news values on the Internet environment rather than a distinct quality. The more elaborate characterisation of social news in relation to traditional news indicates that it involves a different positioning of the author, addressing communication to a specific social group, a different selection of topics than in traditional news, an open critique of traditional journalism and journalistic conventions, and, last but not least, an emphasis on expressing feelings and emotions, which translates into the form: 'social news content is fun to share amongst peers. It reflects, and is embedded in, the emotive and playful sociable communicative practices already prevalent on these platforms (Hurcombe et al., 2021: 402). The case is indeed interesting: social news seems to be a different genre, quite distant from news, although, as some authors point out, nevertheless attached to elements of the journalistic ethos such as objectivity or factual reporting (Hurcombe et al., 2021: 400). The issue certainly requires further research.

In the context of this discussion, the most pertinent question is to what extent the observations on digital modifications of news or the possible emergence of a new online genre (social news) has implications for the study of fake news as a genre. Circulating mainly online, fake news certainly adopts the modifications attributed to online news. However, if these modifications, as discussed, are not fundamental to the description and in-depth analysis of the news genre in general, neither do they undermine the sense of using this category in describing fake news. Contrary to the apparent similarities (for example, shareability), fake news also appears to be distant from social news, as has been characterised above. Of course, some of the fake news is to some extent memetic, sometimes ridiculous and emotional, but it seems that the jocularity, a certain lack of seriousness and a kind of distance characteristic of social news, as well as modifications in terms of subject matter, are not characteristic features of fake news. This observation confirms the original assumption of this chapter—fake news builds its communicative characteristics by reference to journalistic news and therefore represents its genre characteristics. These, in turn, are different, as mentioned earlier, from the genre characteristics of social news.

References

Aitchison, J., & Lewis, D. (Eds.). (2003). *New media language* (1st ed.). Routledge.

Allan, S. (2006). *Online news: Journalism and the Internet*. McGraw-Hill Education.

Bakhtin, M. M. (2010). *The dialogic imagination: Four essays*. University of Texas Press.

Baptista, J. P., & Gradim, A. (2020). Understanding fake news consumption: A review. *Social Sciences, 9*(10), 185.

Bazerman, C. (2013). Genre as social action. In *The Routledge handbook of discourse analysis* (pp. 226–238). Routledge.

Bekalu, M. A. (2006). Presupposition in news discourse. *Discourse & Society, 17*(2), 147–172.

Belcher, D. D. (2023). Digital genres: What they are, what they do, and why we need to better understand them. *English for Specific Purposes, 70*, 33–43.

Bell, A. (1991). *The language of news media* (1. publ). Blackwell.

Bell, A. (1994). Telling stories. In D. Graddol & O. Boyd-Barrett (Eds.), *Media texts, authors and readers: A reader*. Multilingual Matters.

Bhatia, V. K. (2014). *Analysing genre: Language use in professional settings*. Routledge.

Bhatia, V. K. (2014). *Worlds of written discourse. A genre-based view*. Bloomsbury.

Bhatia, V. K. (2014a). Integrating products, processes, purposes and participants in professional writing. In *Writing: Texts, processes and practices* (pp. 21–39). Routledge.

Burgess, J. (2006). Hearing ordinary voices: Cultural studies, vernacular creativity and digital storytelling. *Continuum, 20*(2), 201–214.

Buvarp, P. M. (2023). The space of influence: Developing a new method to conceptualise foreign information manipulation and interference on social media. *Journal of Information Warfare 2023, 22*(2), 31–51.

Connor, U., & Mauranen, A. (1999). Linguistic analysis of grant proposals: European Union research grants. *English for Specific Purposes, 18*(1), 47–62.

Cortada, J. W., & Aspray, W. (2019). *Fake news nation: The long history of lies and misinterpretations in America*. Rowman & Littlefield.

D'Angelo, P., & Kuypers, J. A. (Eds.). (2009). *Doing news framing analysis: Empirical and theoretical perspectives* (1st ed.). Routledge.
Dai, Z., Taneja, H., & Huang, R. (2018, August). Fine-grained structure-based news genre categorization. In *Proceedings of the Workshop Events and Stories in the News 2018* (pp. 61–67).
De Vreese, C. H. (2005). News framing: Theory and typology. *Information Design Journal + Document Design, 13*(1), 51–62.
Deuze, M. (2005). What is journalism? Professional identity and ideology of journalists reconsidered. *Journalism, 6*(4), 442–464.
Egelhofer, J. L., & Lecheler, S. (2019). Fake news as a two-dimensional phenomenon: A framework and research agenda. *Annals of the International Communication Association, 43*(2), 97–116.
Eriksen, L. B., & Ihlström, C. (2000). Evolution of the web news genre: The slow move beyond the print metaphor. In *Proceedings of the 33rd Hawaii International Conference on System Sciences (HICSS)*, Maui, Hawaii, January 2000.
Erjavec, K., & Poler Kovačič, M. (2009). A discursive approach to genre: Mobi news. *European Journal of Communication, 24*(2), 147–164.
Fairclough, N. (1992). Discourse and social change. *Polity.*
Fairclough, N. (1995). *Media discourse.* Arnold.
Fairclough, N. (2000). *New labour, new language?* (1st ed.). Routledge.
Fairclough, N. (2001). The Discourse of New Labour: Critical Discourse Analysis. In M. Wetherell, S. Taylor, & S. J. Yates (Eds.), *Discourse as Data. A Giude for analysis.* SAGE.
Fairclough, N. (2003). *Analysing discourse: textual analysis for social research.* Routledge.
Feuer, J. (1992). Genre study and television. *Channels of Discourse, Reassembled: Television and Contemporary Criticism, 2*, 104–121.
Fiske, J., & Hartley, J. (2004). *Reading television.* Routledge.
Foucault, M. (1972). *The archaeology of knowledge and the discourse on language.* Trans. from the French by A. M. Sheridan Smith. Pantheon Books.
Fras, J. (2013). *O typologii wypowiedzi medialnych i dziennikarskich.* Wydawn. Uniw. Wrocławskiego.
Gibbs, M., Meese, J., Arnold, M., Nansen, B., & Carter, M. (2014). #Funeral and Instagram: Death, social media, and platform vernacular. *Information, Communication & Society, 18*(3), 255–268.
Goffman, E. (1974). Frame analysis: An essay on the organization of experience. *Harrperand Row.*
Greimas, A. J. (1971). Narrative grammar: Units and levels. *MLN, 86*(6), 793–806.
Gruber, H. (2008). 16. Specific genre features of new mass media. *Handbook of Communication in the Public Sphere, 4*, 363.
Halliday, M. (1978). *Language as social semiotics.* Edward Arnold.

Hartley, J. (2013). *Understanding news*. Routledge.

Hermida, A. (2010). Twittering the news: The emergence of ambient journalism. *Journalism Practice, 4*(3), 37–41.

Hodge, B. (1979). Newspapers and communities. In R. Fowler et al. (Eds.), *Language and control*. Routledge & Kegan Paul.

Horne, B., & Adali, S. (2017, May). This just in: Fake news packs a lot in title, uses simpler, repetitive content in text body, more similar to satire than real news. In *Proceedings of the international AAAI conference on web and social media* (Vol. 11, No. 1, pp. 759–766).

Hurcombe, E., Burgess, J., & Harrington, S. (2021). What's newsworthy about 'social news'? Characteristics and potential of an emerging genre. *Journalism, 22*(2), 378–394.

Jenkins, H., & Deuze, M. (2008). Editorial: Convergence culture. *Convergence, 14*(1), 5–12.

Juez, L. A., & Mackenzie, J. L. (2019). Emotion, lies, and "bullshit" in journalistic discourse. *Ibérica, 38*, 17–50.

Khan, A., Brohman, K., & Addas, S. (2022). The anatomy of 'fake news': Studying false messages as digital objects. *Journal of Information Technology, 37*(2), 122–143.

Kopytowska, M. (2015). Covering conflict: Between universality and cultural specificity in news discourse, genre and journalistic style. *International Review of Pragmatics, 7*(2), 308–339.

Kress, G. (2010). *Multimodality: A social semiotic approach to contemporary communication*. Routledge.

Kress, G., & Van Leeuwen, T. (2001). *Multimodal discourse: The modes and media of contemporary communication*. Arnold Publishers.

Laclau, E., & Mouffe, C. (1985). *Hegemony and socialist strategy: Towards a radical democratic politics*. Verso.

Lazer, D. M., Baum, M. A., Benkler, Y., Berinsky, A. J., Greenhill, K. M., Menczer, F., et al. (2018). The science of fake news. *Science, 359*(6380), 1094–1096.

Longacre, R. E. (1974). Narrative versus other discourse genre. In R. Brend (Ed.), *Advances in tagmemics* (pp. 357–376). North Holland.

Maci, S., Demata, M., McGlashan, M., & Seargeant, P. (Eds.). (2024). *The Routledge handbook of discourse and disinformation*. Routledge.

Majid Akhtar, M., Masood, R., Ikram, M., & Kanhere, S. S. (2023). False information, bots and malicious campaigns: Demystifying elements of social media manipulations. *arXiv e-prints*, arXiv-2308.

Martin, J., & White, P. R. (2005). *The language of evaluation*. Palgrave Macmillan.

Martin, J. R., Christie, F., & Rothery, J. (1987). Social processes in education: A reply to Sawyer and Watson (and others). *The Place of Genre in Learning: Current Debates, 3545*.

Miller, C. R. (1984). Genre as social action. *The Quarterly Journal of Speech, 70*(2), 151–167.
Neale, S. (1990). Questions of genre. *Screen, 31*(1), 45–66.
Orlikowski, W. J., & Yates, J. (1994). Genre repertoire: The structuring of communicative practices in organizations. *Administrative Science Quarterly, 39*(4), 541–574.
Pan, Z., & Kosicki, G. M. (1993). Framing analysis: An approach to news discourse. *Political Communication, 10*(1), 55–75.
Paul, N. (1995). Content: A re-visioning. Production in the electronic products newsroom. : The Poynter Institute.
Pavlik, J. V. (2001). *Journalism and new media*. Columbia University Press.
Propp, V. (1968). *Morphology of the folktale*. University of Texas Press.
Richardson, J. E. (2009). *Analysing newspapers: An approach from critical discourse analysis*. Palgrave Macmillan.
Schäfer, S. (2023). Incidental news exposure in a digital media environment: A scoping review of recent research. *Annals of the International Communication Association, 47*(2), 242–260.
Schudson, M. (1978). *Discovering the news: A social history of American newspapers*. Basic Books.
Sheller, M. (2015). News now: Interface, ambience, flow, and the disruptive spatio-temporalities of mobile news media. *Journalism Studies, 16*(1), 12–26.
Shu, K., & Liu, H. (2022). *Detecting fake news on social media*. Springer Nature.
Song, Y., & Lin, A. (2012). A genre-aware approach to online journalism education. *Procedia—Social and Behavioral Sciences, 51*, 400–404.
Stasiuk-Krajewska, K. (2018). *Dziennikarstwo i media. Struktury dyskursu i hegemonia*. Elipsa.
Stasiuk-Krajewska, K. (2024). *Fake news – ujęcie dyskursywne*. Rekonesans.
Swales, J. M. (1990). *Genre analysis: English in academic and research settings*. Cambridge University Press.
Swales, J. M. (2009). Worlds of genre-metaphors of genre. *Genre in a Changing World, 8*(1), 3–16.
Tandoc, E. C., & Jenkins, J. (2017). The Buzzfeedication of journalism? How traditional news organizations are talking about a new entrant to the journalistic field will surprise you! *Journalism, 18*(4), 482–500.
Trew, T. (1979a). Theory and ideology at work. In R. Fowler et al. (Eds.), *Language and Control*. Routledge & Kegan Paul.
Trew, T. (1979b). What the papers say: Linguistic variation and ideological difference. In R. Fowler et al. (Eds.), *Language and control*. Routledge & Kegan Paul.
van Dijk, T. A. (1983). Discourse analysis: Its development and application to the structure of news. *Journal of Communication, 33*(2), 20–43.
van Dijk, T. A. (1985). Structures of news in the press. *Discourse and Communication: New Approaches to the Analysis of Mass Media Discourse and Communication, 10*, 69.

van Dijk, T. A. (1986). News schemata. In S. Greenbaum (Ed.), *Studying writting. Linguistic approaches.* SAGE Publications.
van Dijk, T. A. (1987). *News analysis: Case studies of international and national news in the press* (1st ed.). Routledge.
van Dijk, T. A. (1988). *News as discourse* (1st ed.). Routledge.
van Dijk, T. A. (1989). New developments in discourse analysis (1978–1988). *Journal of Interdisciplinary Literary Studies, 1*, 119–145.
van Dijk, T. A. (1991a). *Racism and the press* (1st ed.). Routledge.
van Dijk, T. A. (1991b). Media contents. The interdisciplinaty study of a news discourse. In N. W. Jankowski & K. B. Jensen (Eds.), *A handbook of qualitative methodologies for mass communication research.* Routledge.
van Dijk, T. A. (2019). *Macrostructures: An interdisciplinary study of global structures in discourse, interaction, and cognition.* Routledge.
Wahl-Jorgensen, K., & Hanitzsch, T. (Eds.). (2020). *The handbook of journalism studies* (2nd ed.). Routledge.
Wardle, C., & Derakhshan, H. (2017). *Information disorder: Toward an interdisciplinary framework for research and policymaking* (Vol. 27, pp. 1–107). Council of Europe.
White, P. (1997). Death, disruption and the moral order: The narrative impulse in mass-media hard news reporting. *Genres and Institutions: Social Processes in the Workplace and School, 101*, 133.
White, P. (1998). *Telling media tales: The news story as rhetoric.* Department of Linguistics, Faculty of Arts, University of Sydney.
Wilkoń, A. (2002). *Spójność i struktura tekstu: wstęp do lingwistyki tekstu.* Universitas.
Wojtak, M. (2004). *Gatunki prasowe, wyd.* UMCS.
Woźniak, A. (2010). Gatunki w sieci. In K. Wolny-Zmorzyński & W. Furman (Eds.), *Internetowe gatunki dziennikarskie.* Wydawnictwa Akademickie i Profesjonalne.
Wu, L. (2016). Did you get the buzz? Are digital native media becoming mainstream? *The Official Research Journal of ISOJ, 6.*
Xia, S. A. (2020). Genre analysis in the digital era: Developments and challenges. *ESP Today, 8*(1), 141–159.
Yarlott, W. V., Cornelio, C., Gao, T., & Finlayson, M. (2018, August). Identifying the discourse function of news article paragraphs. In *Proceedings of the Workshop Events and Stories in the News 2018* (pp. 25–33).
Yates, J., & Orlikowski, W. J. (1992). Genres of organizational communication: A structurational approach to studying communication and media. *Academy of Management Review, 17*(2), 299–326.
Zelizer, B. (2004). Language studies and journalism. In B. Zelizer (Ed.), *Taking journalism seriously: News and the academy.* Sage Publications.

CHAPTER 3

Psychological Dimensions of Disinformation: Etiological Factors and Mitigation Strategies

Jakub Kuś

3.1 What Leads to Disinformation?

To illustrate the circumstances under which we become particularly susceptible to disinformation, consider a scenario where an individual experiences a stimulus that elicits intense emotional arousal, primarily manifesting as fear, anxiety, or anger. This stimulus could originate from global or national political events, personal relationships, or self-related occurrences. These affective states are evolutionarily designed to mobilize the individual for defensive action in a broad sense (Kragel & LaBar, 2016; LeDoux, 2017; Kossowska et al., 2020). Consequently, the initial response will be an attempt to regulate these emotions.

In instances where an individual encounters a situation that induces significant emotional activation but simultaneously presents an information deficit—particularly regarding coping mechanisms—the natural cognitive response is to seek out the missing information. This information-seeking behavior typically manifests in the utilization of digital search engines via computers or mobile devices. The individual engages

J. Kuś (✉)
Faculty of Psychology, SWPS University, Wrocław, Poland
e-mail: jkus@swps.edu.pl

M. Wenzel, K. Stasiuk-Krajewska (eds.), *Disinformation in Central Europe*, https://doi.org/10.1007/978-3-031-76871-2_3

in exploratory behavior across news platforms and social media networks, with the objective of acquiring knowledge about the emotionally salient situation through various means.

This phenomenon can be exemplified by an internet user researching developments on the Ukrainian conflict or a parent actively seeking medical interventions for a critically ill child. These examples illustrate the intersection of emotional arousal, information deficits, and subsequent information-seeking behavior in the digital age.

When an individual engages in information-seeking behavior on the internet regarding a salient topic, particularly one of current relevance, they are confronted with an overwhelming volume of search results, often numbering in the hundreds of thousands or more. The individual is exposed to a plethora of headlines from innumerable web sources, each potentially containing pertinent information. Additionally, they encounter a vast array of user-generated content on social media platforms such as X or Facebook, including posts, comments, and specialized groups purporting to consist of domain experts or interested parties.

It is crucial to note that the individual is operating under conditions of high emotional valence and arousal, with a perceived urgency to access and process a maximum number of information sources in minimal time. Kossowska et al. (2020) emphasize that intense emotions triggered by perceived threats and associated with elevated arousal levels promote selective information processing tendencies. In essence, when confronted with emotionally charged or threatening stimuli, our cognitive processes undergo alteration.

3.1.1 Information Stress and Disinformation

In today's digital age, the proliferation of information has led to a phenomenon known as information stress, where individuals experience psychological stress due to an overload of information (Oniszczenko & Ledzińska, 2019). This excess of information can overwhelm individuals, leading to feelings of anxiety and confusion, and an inability to process the vast amount of data available. On the other hand, disinformation, which refers to false information deliberately created to deceive its recipients, has emerged as a significant cybersecurity concern (French et al., 2023). The intentional spread of disinformation poses a threat to individuals, societies, and even democratic institutions, as it can manipulate perceptions, sow discord, and erode trust in reliable sources of information.

Metaphorically, the experience of information stress can be likened to navigating an unfamiliar, complex intersection devoid of traffic signage, with the imperative to traverse it efficiently and effectively. This analogy illustrates the multitude of incoming stimuli: visual (pedestrians, advertising), auditory (emergency signals, vehicular noise), and cognitive (decision-making under time pressure). The resultant perceptual and cognitive overload may lead to attentional failures regarding critical environmental cues. Information stress impairs the cognitive system's ability to differentiate between relevant and irrelevant information. When the influx of informational stimuli exceeds the brain's processing capacity, it may cease to effectively encode and process subsequent, potentially significant, data. The prevalence of conflicting information sources can exacerbate information-induced stress, especially when individuals are faced with contradictory data (Baalen, 2014).

The volume of information and disinformation encountered in modern environments often surpasses cognitive processing capabilities. This overload triggers physiological stress responses, altering normal cognitive functioning. Primary effects include rapid depletion of cognitive resources and increased fatigue. Consequently, decision-making processes and critical thinking abilities are significantly compromised. Attentional focus becomes increasingly difficult to maintain in the presence of multiple, competing information sources. Furthermore, the capacity to discriminate between veracious and false information, as well as between information and disinformation, is substantially impaired.

The experience of information stress serves as a potent mediator in the acceptance of false or misleading information, potentially leading to significant distortions in an individual's perception of reality. This cognitive state increases susceptibility to deception and manipulation, while simultaneously impairing the ability to discriminate between salient and irrelevant information.

The term "junk news" has gained prominence in information science discourse, drawing parallels with the concept of "junk food" in nutritional science. This analogy suggests that certain types of information, while readily consumed, provide minimal cognitive value and may deplete cognitive resources (e.g., news regarding celebrities' personal lives). The contemporary digital landscape increasingly resembles an information ecosystem characterized by high entropy, potentially disorienting users who lack effective information navigation strategies.

The phenomenon of information stress is exacerbated by engagement in multitasking behavior. Turkle (2011) posits that many digital users exhibit a preference for concurrent task execution, potentially falling prey to what is termed the "multitasking illusion" This cognitive bias is characterized by a strong subjective belief in enhanced productivity through simultaneous task performance. However, Turkle emphasizes that this perception is illusory, as compulsive multitaskers typically demonstrate significantly reduced task efficiency compared to sequential task completion.

Popławska et al. (2021) suggest that developed self-regulation competencies may serve as a protective factor against the negative consequences of multitasking. Individuals with low self-regulation capacities may find multitasking appealing and subjectively enjoyable, but this behavior is likely to result in decreased cognitive efficiency and increased information stress. This cognitive state, in turn, may heighten vulnerability to disinformation.

A distinct manifestation of information stress can arise from information scarcity, rather than overabundance. This phenomenon occurs when an individual experiences a strong drive for specific information acquisition but encounters significant barriers to access. Such a situation induces a stress response coupled with intense frustration. Under these conditions of information deprivation, individuals may exhibit heightened credulity towards any available informational cues, regardless of their veracity or reliability. This cognitive state significantly increases susceptibility to disinformation.

The psychological state induced by information scarcity is characterized by uncertainty and ambiguity, which in turn enhances vulnerability to various forms of cognitive manipulation. These manipulative strategies may be designed to increase the perceived credibility of false or misleading information. This scenario presents a notable risk factor for falling victim to disinformation campaigns or inadvertently accepting false narratives. The cognitive discomfort associated with information deprivation may lead individuals to accept lower standards of evidence or bypass normal critical thinking processes in their quest for closure or understanding.

The interplay between information scarcity, cognitive stress, and susceptibility to disinformation underscores the complex relationship between information availability, cognitive processing, and belief formation. This phenomenon highlights the need for robust information literacy skills and critical thinking abilities, particularly in environments characterized by limited or restricted information access.

In the contemporary era characterized by pervasive disinformation, the concept of illusory truth effect through repetition has resurfaced with renewed relevance. Unkelbach and Speckmann (2021) empirically validated this phenomenon, demonstrating that repetition particularly enhances the perceived veracity of information that is initially unfamiliar and lacks widespread dissemination. This finding has significant implications for the propagation of disinformation, as novel or obscure false narratives may gain credibility through repeated exposure, especially in social media ecosystems.

Differential cognitive processing of veridical and false information has been observed in online user behavior. Barfar (2019) has demonstrated that exposure to political disinformation, compared to reliable information, results in reduced analytical complexity in user-generated responses. This suggests that consumers of false information engage in less reflective cognitive processing. Furthermore, disinformation elicits heightened emotional reactivity, particularly in terms of anger and hostility. These findings indicate that exposure to disinformation negatively impacts critical thinking capabilities, increases impulsivity, and intensifies emotional responses.

Psychological research on disinformation has identified specific cognitive factors that exacerbate susceptibility to false information in public discourse. Effron and Helgason (2022) highlight three primary factors: partisanship, imagination, and repetition. In the post-truth era, creators of disinformation may not necessarily aim to convince the public of the veracity of their claims. Instead, they may appeal to in-group affiliations or partisan biases, which can lead to a lowering of moral standards in evaluating false information, particularly when it aligns with pre-existing beliefs or serves perceived interests.

Effron and Helgason (2022) also highlight the existence of a mechanism that appeals to the imagination. It involves relying on disinformation campaigns to convince audiences that a given lie may have been true or, particularly importantly, may become so in the future. The third factor, repetition, is the most basic but unfortunately very effective. It refers to the old adage that a lie repeated a thousand times becomes the truth. "Becomes", of course, in a purely subjective sense, in terms of an individual's perception of a statement as true. Such repetition lowers the likelihood of moral condemnation of a false message, using the mechanism of 'stripping' it of its emotional charge. In turn, if something does not evoke emotions in us (e.g. shame, anger, guilt), then the need to apologise for it will not arise. Effron and Helgason (2022, p. 4) very aptly conclude that we do not live in times so filled with disinformation and various kinds of fake news because people lie, and others believe these lies. We live in such times because it is worryingly easy to get a 'moral pass' on dishonesty, even when others know you are lying.

3.1.2 Personality Traits and Disinformation

Recent psychological research has increasingly focused on the potential association between specific personality traits and individual susceptibility or resilience to disinformation. Calvillo et al. (2021) utilized the widely accepted Big Five personality model (Pervin & Cervone, 2022), which postulates five fundamental personality dimensions: extraversion, neuroticism, openness to experience, agreeableness, and conscientiousness. These dimensions are operationalized as follows (Pervin & Cervone, 2022):

1. Extraversion: Characterized by sociability, loquaciousness, ease of forming new social connections, propensity for enjoyment, and pursuit of intense external stimuli.
2. Neuroticism: Manifested through frequent experiences of negative affect (e.g., sadness, anger, guilt), emotional instability, irrational behavioral tendencies, and maladaptive stress-coping mechanisms.
3. Openness to experience: Exhibited through a persistent pursuit of novel experiences, cognitive curiosity, and tolerance for ambiguity and uncertainty.
4. Agreeableness: Defined by a positive, trusting interpersonal orientation, altruistic tendencies, and cooperative disposition.
5. Conscientiousness: Reflected in work commitment, goal persistence, efficiency motivation, and general meticulousness in life domains.

Recent empirical investigations have elucidated the complex interplay between personality traits and an individual's capacity to discern veracious information from disinformation. Calvillo et al. (2021) identified a positive correlation between conscientiousness and openness to experience with the ability to differentiate between factual and fabricated news. Conversely, this discriminatory aptitude was negatively associated with extraversion and time allocated to online news consumption. These findings were corroborated by Wolverton and Stevens (2019), who observed analogous relationships between personality dimensions and disinformation detection capabilities.

A seminal study in this domain, conducted by Miller et al. (2023), explored the intricate relationships between personality traits and media truth discernment, operationalized through three constructs: critical media use, functional media literacy, and generalized communicative suspicion. Utilizing the Big Five personality framework, the researchers discovered positive associations between both conscientiousness and agreeableness with critical media literacy. Additionally, openness to experience and conscientiousness demonstrated positive correlations with functional media use. The authors posit that conscientiousness emerges as the most salient trait in this context, potentially serving as both a protective and risk factor for vulnerability to disinformation.

Miller et al. (2023) emphasize also the nuanced interplay between conscientiousness and agreeableness. Their research indicates that individuals who score highly on both traits may recognize disinformation as an alternative perspective in an argument, rather than outright falsehood. Elevated levels of agreeableness tend to foster credulity and a presumption of benevolent intent in others. Consequently, the combination of agreeableness and conscientiousness can manifest in a strong desire to consider all perspectives on an issue, while simultaneously increasing the risk of legitimizing false narratives as valid viewpoints in political discourse.

In conclusion, these empirical investigations underscore the multifaceted relationship between personality traits and susceptibility to disinformation. While certain trait combinations may predispose individuals to heightened vulnerability, emerging research indicates that targeted interventions and educational initiatives can play a crucial role in enhancing critical thinking skills and media literacy across diverse personality profiles. The most effective strategy for countering the spread of disinformation in modern information environments may involve a comprehensive approach that integrates personality evaluation, cognitive training, and sociocultural awareness.

3.2 The Impact of Disinformation on the Functioning of Societies

Disinformation, characterized by false or misleading information disseminated with the intent to deceive or for economic gain, has far-reaching implications across various sectors, including the economy, politics, and public health. The COVID-19 pandemic has served as fertile ground for the spread of disinformation, taking advantage of existing political and epistemological divides to contest policy responses (Mascio et al., 2021). This dissemination of disinformation has been particularly effective in polarized political environments, where partisan actors use disinformation to undermine expert knowledge and promote mistrust. The consequences of disinformation extend to delaying societal development, for example as seen in the case of the negative attitude towards 5G technologies in Poland (Ejdys, 2023). The significance of accurate and informative content dissemination by authoritative bodies like the World Health Organization (WHO) during global health crises has been underscored to combat the spread of disinformation.

Empirical studies have revealed alarming statistics regarding the prevalence of disinformation. Research indicates that as early as 2016, over 70% of news content shared on social media platforms consisted of disinformation (Miller et al., 2024). This proportion is hypothesized to have increased in subsequent years. Moreover, studies suggest that disinformation exhibits a higher rate of propagation and broader reach in online environments compared to factual information, a phenomenon attributed to its accessibility and emotive content (Miller et al., 2024; Vosoughi et al., 2018). The spread of disinformation on social media has become a pressing challenge globally, necessitating legal responses to combat its harmful effects (Huang, 2024). The ease of access to social media platforms and the rapid dissemination of information make it crucial to address the spread of false information effectively.

In the realm of political science and communication studies, disinformation has emerged as a critical factor in the formation of public opinion, the determination of electoral outcomes, and the exacerbation of societal polarization. The proliferation of disinformation in digital environments has been observed to significantly impact democratic discourse and pose substantial risks to public health and safety (Vosoughi et al., 2018). Exposure to false or misleading information pertaining to pivotal issues, such as global health crises (e.g., the COVID-19 pandemic) and national electoral processes, has been associated with divergent perceptions of social, economic, and political phenomena. This cognitive divergence contributes to increased polarization, ideological radicalization, and, in extreme cases, the manifestation of violent behavior at both individual and collective levels (Guess et al., 2020).

The dissemination of erroneous information has been demonstrated to exert significant influence on public sentiment, shape policy formulation processes, and potentially compromise the integrity of democratic institutions (Allcott & Gentzkow, 2017). Recent research by Pennycook and Rand (2021) suggests that the spread of disinformation is facilitated by cognitive factors such as analytical thinking deficits and partisan motivated reasoning, rather than solely by exposure to false content.

The multifaceted nature of disinformation's impact on political processes necessitates interdisciplinary approaches to its study and mitigation. Scholars in fields ranging from political psychology to network science are increasingly focusing on developing robust methodologies for detecting, analyzing, and countering disinformation in various contexts (Lazer et al., 2018). As digital platforms continue to evolve, understanding the dynamics of disinformation dissemination and its effects on democratic systems remains a critical area of ongoing research.

Moreover, the impact of disinformation extends beyond individual beliefs and behavior to systemic issues such as the dissemination of fake news and the manipulation of public discourse. Disinformation distorts reality and hinders effective governance and policymaking. Strategic communication initiatives aimed at correcting disinformation often employ rigorous experimental methods to evaluate the effectiveness of various correction strategies. Fact-checking plays a vital role in combating disinformation by providing accurate information to counter false narratives and misleading claims (York et al., 2020). However, the impact of fact-checking efforts can be influenced by individuals' political attitudes and cognitive biases, highlighting the complex relationship between disinformation and belief systems (Ecker & Ang, 2019).

The ramifications of disinformation extend far beyond immediate political or health consequences, significantly impacting social innovation and economic potential across nations. Vasist and Krishnan's (2023a) research posits a correlation between the entrenchment of disinformation in political discourse and a diminution of societal innovative capacity. They argue that an environment saturated with false information and cynical manipulation serves as a significant impediment to creativity and progress.

Furthermore, they postulate that persistent exposure to socio-political disinformation engenders a state of perpetual conflict, which consequently erodes individuals' motivation to innovate.

Researchers in the field of information studies have proposed the conceptual framework of 'informational pollution' to elucidate the ubiquitous nature of disinformation and disinformation in contemporary digital environments (Vaidhyanathan, 2018; Bartlett, 2019). This theoretical construct posits that the proliferation of false or misleading information creates a cognitive environment analogous to atmospheric pollution, potentially impeding individual cognitive processes and collective societal advancement. The metaphorical association between disinformation and environmental contaminants effectively illustrates the pervasive and deleterious impact of inaccurate information on digital ecosystems, potentially inducing states of epistemic uncertainty and eroding social trust mechanisms.

The challenge of mitigating disinformation has not gone unnoticed by governing bodies. The European Commission's promulgation of an enhanced Code of Conduct on Combating Disinformation, in conjunction with the Digital Services Act on June 16, 2022, serves as a testament to the severity of this issue and its potential to significantly alter the global socio-political landscape. Interestingly, the impact of disinformation extends beyond the realm of politics and into sectors such as tourism. Research by Vasist and Krishnan (2023b) indicates that potential visitors express reluctance to travel to countries where disinformation is prevalent, citing concerns about political instability.

Empirical investigations have elucidated a significant inverse relationship between disinformation exposure and consumer confidence in marketed products, with subsequent detrimental effects on economic indicators (Balcaen et al., 2023). The ubiquitous nature of disinformation within digital ecosystems, exemplified by the proliferation of falsified product reviews, engenders a challenging environment for consumers to discriminate between veracious and fabricated content.

The propagation of disinformation exhibits characteristics analogous to viral transmission within societal structures, potentially compromising multiple domains of societal functioning. Balcaen et al. (2023) posit that the erosion of trust in a single, salient domain may catalyze a cascading effect, amplifying distrust across various societal sectors if left unmitigated.

Of particular concern is the potential for disinformation to influence behavioral outcomes even in the absence of conscious awareness. Bastick

(2021) demonstrated that brief exposure (<5 minutes) to false information was sufficient to induce significant modifications in unconscious behavior, impacting decision-making processes across consumer, health, and political domains.

Given the documented limitations in individuals' introspective access to higher-order cognitive processes, particularly regarding decision-making determinants (Nisbett & Wilson, 1977), the role of disinformation in shaping unconscious processes warrants further investigation. The potential for covert manipulation through disinformation in domains of personal significance presents a substantial challenge, as behavioral outcomes may be influenced without conscious recognition of the underlying informational manipulation.

Disinformation has emerged as a paramount challenge in contemporary society. Over two decades ago, Sassenberg et al. (2003) presciently identified the comprehensive analysis of novel technologies' impact on human cognition and behavior as a critical objective for modern psychological research.

The term 'infodemia' has gained significant traction in recent academic discourse. It is operationally defined as an experiential state resulting from excessive information exposure, leading to cognitive overload, decision-making impairment, and increased propensity towards risk-taking behavior, particularly in health-related domains.

Psychological researchers investigating this phenomenon have identified numerous significant implications for decision-making processes under conditions of uncertainty. In the context of economic decision-making, disinformation can precipitate market inefficiencies, suboptimal resource allocation, and diminished investor confidence. Addressing disinformation in the economic sphere necessitates targeted interventions aimed at enhancing information literacy, promoting critical thinking skills, and fostering transparency in financial reporting and communication.

Kossowska et al. (2020, p. 15) posit that "prolonged exposure to fear, anxiety, and uncertainty degrades individual health outcomes, diminishes effective action capacity, and engenders debilitating defensive behavior". Functioning within an infodemic environment represents a significant psychological stressor with direct implications for daily functioning. The cognitive burden of distinguishing veracious information from false narratives in an oversaturated information landscape presents tangible risks to individual well-being.

Combating disinformation necessitates a multifaceted approach, encompassing technological solutions such as blockchain-based platforms that leverage collective intelligence and artificial intelligence to identify and counteract false information (Butincu & Alexandrescu, 2023). The implementation of innovative technologies enables enhanced detection and mitigation of disinformation campaigns, thereby contributing to the preservation of information ecosystem integrity. Furthermore, legal responses at both national and international levels are imperative to address the challenges posed by disinformation, particularly in the context of evolving threats to democratic processes and societal stability (Watney, 2023).

In conclusion, the influence of disinformation on the economic, political, and social spheres is multifaceted and far-reaching. From distorting market dynamics and shaping consumer behavior to influencing electoral outcomes and fueling social polarization, disinformation poses a significant challenge to the functioning of democratic societies. Addressing the impact of disinformation requires a comprehensive approach that combines fact-checking initiatives, targeted interventions, and efforts to enhance media literacy and critical thinking skills. By promoting a culture of truth-seeking and evidence-based decision-making, societies can mitigate the harmful effects of disinformation and safeguard the integrity of public discourse and democratic processes.

3.3 Psychological Tools in the Fight Against Disinformation

3.3.1 Verify the Source of Information

Numerous disinformation campaigns exploit the tendency of many internet users to neglect source verification and credibility assessment of online information. Consequently, the primary defense mechanism against disinformation involves rigorous source authentication and evaluation of the information's origin (Amazeen & Bucy, 2019). This process necessitates determining the reliability of the source (e.g., a website, discussion forum, social media profile) and, paradoxically, verifying its actual existence.

The proliferation of disinformation attempts utilizing fabricated senders, such as internet bots or pseudo-websites, underscores the importance of thorough verification (Shao et al., 2018). These artificial entities are often created solely for the purpose of specific manipulative actions, highlighting the need for enhanced digital literacy skills among users.

A substantial body of psychological research emphasizes the crucial role of critical thinking as the most effective cognitive tool for combating disinformation and fake news (Kleiman et al., 2015; Pennycook & Rand, 2019). Critical thinking skills enable individuals to evaluate information systematically, identify logical fallacies, and make evidence-based judgments. In response to the growing threat of disinformation various organizations and foundations have developed educational initiatives aimed at enhancing media literacy and critical thinking skills (Bulger & Davison, 2018). These programs often employ age-specific curricula, utilizing tailored examples to improve disinformation detection methods in alignment with the cognitive capacities of diverse audiences (Jeong et al., 2012).

3.3.2 Examine Your Emotions Closely

Disinformation campaigns frequently employ strategies aimed at eliciting intense emotional responses, potentially compromising cognitive processing depth (Kahneman, 2012). These campaigns predominantly target the activation of strong, negative emotions such as xenophobia or fear (Brady et al., 2017). Decades of psychological research have demonstrated that emotional arousal significantly alters cognitive processes, with the mind adapting to the experienced emotional state (Lerner et al., 2015).

The digital environment presents a particularly conducive platform for emotion-based disinformation dissemination. Visual content, including manipulated images and videos, plays a crucial role in reinforcing false narratives (Nightingale et al., 2017). The creation and distribution of such content have become increasingly accessible, requiring minimal technical expertise. Individual differences in temperament, particularly the propensity for sensation-seeking, may predispose certain individuals to greater susceptibility to disinformation (Zuckerman, 2007). This vulnerability is exacerbated by the social amplification of emotions within group settings, a phenomenon known as emotional contagion (Hatfield et al., 1993).

The concept of groupthink in social psychology (Janis, 1982) elucidates how individuals may adjust their cognitive processes to align with group norms. This conformity can lead to impaired critical thinking and selective information processing, whereby contradictory evidence is filtered out to maintain group cohesion (Sunstein & Hastie, 2015). Disinformation acts as an "emotional steroid," fueling these group processes and reinforcing existing beliefs within information bubbles (Del Vicario et al., 2016). However, metacognitive awareness of one's emotional state and the potential for manipulation can serve as a protective factor against disinformation (Pennycook & Rand, 2019).

The effectiveness of disinformation campaigns often relies on their ability to evoke emotions such as fear, anger, or contempt through coordinated and targeted efforts (Vosoughi et al., 2018). Enhancing emotional intelligence and developing critical media literacy skills may significantly improve individuals' resilience to such manipulative tactics (Jones-Jang et al., 2021).

3.3.3 *Avoid Impulsive Actions*

The role of affective processes in the propagation of disinformation is paramount, particularly when individuals are exposed to emotionally charged content from diverse sources (Martel et al., 2020). A common behavioral response to emotive stimuli in digital environments involves rapid engagement through comments, endorsements, or content sharing. However, in the contemporary information ecosystem, exercising caution against impulsive emotional reactions is crucial, as such behavior may inadvertently contribute to the dissemination of disinformation and perpetuation of false narratives (Pennycook et al., 2021).

A dual-pronged approach to emotional regulation in digital spaces is recommended. Firstly, individuals should engage in metacognitive monitoring of their affective states, identifying situational triggers that elicit intense emotional responses (Gross, 2015). Secondly, the implementation of cognitive reappraisal strategies to mitigate impulsive actions is advisable, encouraging deliberate evaluation before content dissemination or engagement (Aldao et al., 2010).

In the context of information overload, often referred to as "infodemic" (Zarocostas, 2020), it is imperative to cultivate a multifaceted perspective on information processing and to exercise restraint in immediate action. The digital footprint left by online interactions underscores the importance of considered engagement, as each action potentially contributes to broader information cascades (Vosoughi et al., 2018).

It is crucial to critically evaluate the potential consequences of digital actions, even those perceived as insignificant, such as sharing content on social media platforms. This approach aligns with the concept of digital citizenship, emphasizing responsible and ethical online behavior (Choi, 2016).

3.4 Summary

The development of efficacious strategies to counteract disinformation has emerged as one of the most critical challenges in contemporary scientific discourse. This multifaceted issue necessitates intensive interdisciplinary collaboration across the natural sciences, medicine, social sciences, and humanities. We are arguably at the precipice of an era that will redefine the scale and scope of information accessibility challenges. This discourse aims to examine the phenomenon of disinformation through a psychological lens. It is imperative to note that this analysis is predicated on the current state of knowledge, which, given the rapid evolution of internet psychology, may soon be rendered obsolete due to technological advancements. The impending phase of the digital revolution, particularly associated with virtual reality, is poised to introduce novel vectors for disinformation dissemination (Bailenson, 2021).

Presently, numerous organizations and non-governmental entities are raising alarms about emergent threats such as "deepfakes." This phenomenon is conceptualized as a technology leveraging artificial intelligence "to create or manipulate audio-visual content with the intent of depicting events that never transpired" (Dabrowska, 2020, p. 89). The potential for misuse is exemplified by the feasibility of generating synthetic media depicting a head of state declaring war, with high fidelity to their authentic voice and mannerisms (Westerlund, 2019).

In light of these developments, we stand on the threshold of a significant societal transformation. The scientific community faces a formidable interdisciplinary challenge in addressing these emerging threats. Technological solutions alone are not enough; they need to be supported by strong social and educational initiatives to promote critical thinking and digital literacy among diverse communities. As we navigate this new era, it is crucial that research efforts remain agile, interdisciplinary, and responsive to rapidly changing technological and social dynamics. The future of information integrity and societal trust may well depend on our collective ability to rise to this multifaceted challenge.

Bibliography

Aldao, A., Nolen-Hoeksema, S., & Schweizer, S. (2010). Emotion-regulation strategies across psychopathology: A meta-analytic review. *Clinical Psychology Review, 30*(2), 217–237. https://doi.org/10.1016/j.cpr.2009.11.004

Allcott, H., & Gentzkow, M. (2017). Social media and fake news in the 2016 election. *Journal of Economic Perspectives, 31*(2), 211–236. https://doi.org/10.1257/jep.31.2.211

Amazeen, M. A., & Bucy, E. P. (2019). Conferring resistance to digital disinformation: The inoculating influence of procedural news knowledge. *Journal of Broadcasting & Electronic Media, 63*(3), 415–432. https://doi.org/10.1080/08838151.2019.1653101

Baalen, M. v. (2014). Adaptation, conflicting information, and stress. *Biological Theory, 9*(4), 431–439. https://doi.org/10.1007/s13752-014-0184-9

Bailenson, J. N. (2021). Nonverbal overload: A theoretical argument for the causes of Zoom fatigue. *Technology, Mind, and Behavior, 2*(1). https://doi.org/10.1037/tmb0000030

Balcaen, P., Buts, C., Du Bois, C. & Tkacheva O. (2023). The effect of disinformation about COVID-19 on consumer confidence: Insights from a survey experiment. *Journal of Behavioral and Experimental Economics, 102*, 101968. https://doi.org/10.1016/j.socec.2022.101968

Barfar, A. (2019). Cognitive and affective responses to political disinformation in Facebook. *Computers in Human Behavior, 101*, 173–179. https://doi.org/10.1016/j.chb.2019.07.026

Bartlett, J. (2019). *People against technology. How the internet is killing democracy (and how we can save it)*. Post Factum Publishing.

Bastick, Z. (2021). Would you notice if fake news changed your behaviour? An experiment on the unconscious effects of disinformation. *Computers in Human Behavior, 116*, 106633. https://doi.org/10.1016/j.chb.2020.106633

Brady, W. J., Wills, J. A., Jost, J. T., Tucker, J. A., & Van Bavel, J. J. (2017). Emotion shapes the diffusion of moralized content in social networks. *Proceedings of the National Academy of Sciences, 114*(28), 7313–7318. https://doi.org/10.1073/pnas.1618923114

Bulger, M., & Davison, P. (2018). The promises, challenges, and futures of media literacy. *Journal of Media Literacy Education, 10*(1), 1–21. https://doi.org/10.23860/JMLE-2018-10-1-1

Butincu, C., & Alexandrescu, A. (2023). Blockchain-based platform to fight disinformation using crowd wisdom and artificial intelligence. *Applied Sciences, 13*(10), 6088. https://doi.org/10.3390/app13106088

Calvillo, D., Garcia, R., Bertrand, K., & Mayers, T. (2021). Personality factors and self-reported political news consumption predict susceptibility to political fake news. *Personality and Individual Differences, 174*, 110666. https://doi.org/10.1016/j.paid.2021.110666

Choi, M. (2016). A concept analysis of digital citizenship for democratic citizenship education in the internet age. *Theory & Research in Social Education, 44*(4), 565–607. https://doi.org/10.1080/00933104.2016.1210549

Dabrowska, L. (2020). Deep fake – A new dimension of online manipulation. *Media Management, 8*(2), 89-101. https://doi.org/10.4467/23540214ZM.20.024.11803

Del Vicario, M., et al. (2016). The spreading of disinformation online. *Proceedings of the National Academy of Sciences, 113*(3), 554–559. https://doi.org/10.1073/pnas.1517441113

Ecker, U. K. H., & Ang, L. C. (2019). Political attitudes and the processing of disinformation corrections. *Political Psychology, 40*, 241–260. https://doi.org/10.1111/pops.12494

Effron, D., & Helgason, B. (2022). The moral psychology of disinformation: Why we excuse dishonesty in a post-truth world. *Current Opinion in Psychology, 47*, 101375. https://doi.org/10.1016/j.copsyc.2022.101375

Ejdys, J. (2023). The society's attitude toward 5G technologies – A case study of Poland. *Technological and Economic Development of Economy, 29*(5), 1539–1558. https://doi.org/10.3846/tede.2023.20157

French, A., Storey, V. C., & Wallace, L. (2023). A typology of disinformation intentionality and impact. *Information Systems Journal, 34*(4), 1324–1354. https://doi.org/10.1111/isj.12495

Gross, J. J. (2015). Emotion regulation: Current status and future prospects. *Psychological Inquiry, 26*(1), 1–26. https://doi.org/10.1080/1047840X.2014.940781

Guess, A. M., Nyhan, B., & Reifler, J. (2020). Exposure to untrustworthy websites in the 2016 US election. *Nature Human Behaviour, 4*(5), 472–480. https://doi.org/10.1038/s41562-020-0833-x

Hatfield, E., Cacioppo, J. T., & Rapson, R. L. (1993). Emotional contagion. *Current Directions in Psychological Science, 2*(3), 96–100. https://doi.org/10.1111/1467-8721.ep10770953

Huang, J. (2024). Information warfare in the digital age: Legal responses to the spread of false information under public international law. *Journal of Education, Humanities and Social Sciences, 28*, 176–184. https://doi.org/10.54097/46jmtq31

Janis, I. L. (1982). *Groupthink: Psychological studies of policy decisions and fiascoes.* Houghton Mifflin.

Jeong, S. H., Cho, H., & Hwang, Y. (2012). Media literacy interventions: A meta-analytic review. *Journal of Communication, 62*(3), 454–472. https://doi.org/10.1111/j.1460-2466.2012.01643.x

Jones-Jang, S. M., Mortensen, T., & Liu, J. (2021). Does media literacy help identification of fake news? Information literacy helps, but other literacies don't. *American Behavioral Scientist, 65*(2), 371–388. https://doi.org/10.1177/0002764219869406

Kahneman, D. (2012). *Thinking, fast and slow*. Farrar, Straus and Giroux.

Kleiman, T., Sher, N., Elster, A., & Mayo, R. (2015). Accessibility is a matter of trust: Dispositional and contextual distrust blocks accessibility effects. *Cognition, 142*, 333–344. https://doi.org/10.1016/j.cognition.2015.06.001

Kossowska, M., Letki, N., Zaleśkiewicz, T., & Wichary, S. (2020). *Man in the face of a pandemic*. Smak Słowa Publishing.

Kragel, P., & LaBar, K. (2016). Decoding the nature of emotion in the brain. *Trends in Cognitive Sciences, 20*(6), 444–455. https://doi.org/10.1016/j.tics.2016.03.011

Lazer, D. M., et al. (2018). The science of fake news. *Science, 359*(6380), 1094–1096. https://doi.org/10.1126/science.aao2998

LeDoux, J. (2017). *Anxiety*. Copernicus Center Press.

Lerner, J. S., Li, Y., Valdesolo, P., & Kassam, K. S. (2015). Emotion and decision making. *Annual Review of Psychology, 66*, 799–823. https://doi.org/10.1146/annurev-psych-010213-115043

Martel, C., Pennycook, G., & Rand, D. G. (2020). Reliance on emotion promotes belief in fake news. *Cognitive Research: Principles and Implications, 5*(1), 1–20. https://doi.org/10.1186/s41235-020-00252-3

Mascio, F., Natalini, A., Barbieri, M., & Selva, D. (2021). The role of regulatory agencies in agenda-setting processes: Insights from the Italian response to the covid-19 infodemic. *Swiss Political Science Review, 27*(2), 271–282. https://doi.org/10.1111/spsr.12465

Miller, C. H., Menard, P., & Bourrie, D. M. (2023). Personality traits as predictors of media truth discernment. *Computers in Human Behavior, 138*, 107393. https://doi.org/10.1016/j.chb.2022.107393

Miller, S., Menard, P., & Bourrie, D. (2024). I'm not fluent: How linguistic fluency, new media literacy, and personality traits influence fake news engagement behavior on social media. *Information & Management, 61*(2). https://doi.org/10.1016/j.im.2023.103912

Nightingale, S. J., Wade, K. A., & Watson, D. G. (2017). Can people identify original and manipulated photos of real-world scenes? *Cognitive Research: Principles and Implications, 2*(1), 30. https://doi.org/10.1186/s41235-017-0067-2

Nisbett, R. E., & Wilson, T. D. (1977). Telling more than we can know: Verbal reports on mental processes. *Psychological Review, 84*(3), 231–259. https://doi.org/10.1037/0033-295X.84.3.231

Oniszczenko, W., & Ledzińska, M. (2019). Sex, affective temperaments and information stress. *International Journal of Occupational Medicine and*

Environmental Health, 32(5), 635–644. https://doi.org/10.13075/ijomeh.1896.01392

Pennycook, G., & Rand, D. G. (2019). Lazy, not biased: Susceptibility to partisan fake news is better explained by lack of reasoning than by motivated reasoning. *Cognition, 188*, 39–50. https://doi.org/10.1016/j.cognition.2018.06.011

Pennycook, G., & Rand, D. G. (2021). The psychology of fake news. *Trends in Cognitive Sciences, 25*(5), 388–402. https://doi.org/10.1016/j.tics.2021.02.007

Pennycook, G., Epstein, Z., Mosleh, M., Arechar, A. A., Eckles, D., & Rand, D. G. (2021). Shifting attention to accuracy can reduce disinformation online. *Nature, 592*(7855), 590–595. https://doi.org/10.1038/s41586-021-03344-2

Pervin, L., & Cervone, D. (2022). *Personality. Theory and research.* Wiley.

Popławska, A., Szumowska, E., & Kuś, J. (2021). Why do we need media multitasking? A self-regulatory perspective. *Frontiers in Psychology, 12*, 624649. https://doi.org/10.3389/fpsyg.2021.624649

Sassenberg, K., Boos, M., Postmes, T., & Reips, U.-D. (2003). Studying the Internet: A challenge for modern psychology. *Swiss Journal of Psychology, 62*(2), 75–77. https://doi.org/10.1024/1421-0185.62.2.75

Shao, C., Ciampaglia, G. L., Varol, O., Yang, K. C., Flammini, A., & Menczer, F. (2018). The spread of low-credibility content by social bots. *Nature Communications, 9*(1), 4787. https://doi.org/10.1038/s41467-018-06930-7

Sunstein, C. R., & Hastie, R. (2015). *Wiser: Getting beyond groupthink to make groups smarter.* Harvard Business Press.

Turkle, S. (2011). *Alone together: Why we expect more from technology and less from each other.* Basic Books.

Unkelbach, C., & Speckmann, F. (2021). Mere repetition increases belief in factually true COVID-19-related information. *Journal of Applied Research in Memory and Cognition, 10*(2), 241–247. https://doi.org/10.1016/j.jarmac.2021.02.001

Vaidhyanathan, S. (2018). *Antisocial media. How Facebook is distancing us from each other and threatening democracy.* W.A.B. Publishing.

Vasist, P., & Krishnan, S. (2023a). Powered by innovation, derailed by disinformation: A multi-country analysis of the influence of online political disinformation on nations' innovation performance. *Technological Forecasting and Social Change, 199*, 123029. https://doi.org/10.1016/j.techfore.2023.123029

Vasist, P., & Krishnan, S. (2023b). Disinformation 'gatecrashes' tourism: An empirical study. *Annals of Tourism Research, 101*, 103575. https://doi.org/10.1016/j.annals.2023.103575

Vosoughi, S., Roy, D., & Aral, S. (2018). The spread of true and false news online. *Science, 359*(6380), 1146–1151. https://doi.org/10.1126/science.aap9559

Watney, M. (2023). Legal response to social media disinformation on national level. *European Conference on Cyber Warfare and Security, 22*(1), 525–532. https://doi.org/10.34190/eccws.22.1.1106

Westerlund, M. (2019). The emergence of deepfake technology: A review. *Technology Innovation Management Review, 9*(11), 39–52. https://doi.org/10.22215/timreview/1282

Wolverton, C., & Stevens, D. (2019). The impact of personality in recognizing disinformation. *Online Information Review, 44*(1), 181–191. https://doi.org/10.1108/OIR-04-2019-0115/

York, C., Ponder, J. D., Humphries, Z., Goodall, C., Beam, M., & Winters, C. (2020). Effects of fact-checking political misinformation on perceptual accuracy and epistemic political efficacy. *Journalism & Mass Communication Quarterly, 97*(4), 958–980. https://doi.org/10.1177/1077699019890119

Zarocostas, J. (2020). How to fight an infodemic. *The Lancet, 395*(10225), 676. https://doi.org/10.1016/S0140-6736(20)30461-X

Zuckerman, M. (2007). *Sensation seeking and risky behavior.* American Psychological Association.

CHAPTER 4

Strategies of Disinformation in Central Europe: Narratives Related to the COVID-19 Pandemic and War in Ukraine

Kateřina Turková and *Veronika Macková*

4.1 Introduction

The dissemination of disinformation is a critical issue worldwide. As Gregor and Mlejnková (2021c, p. 541) argue: "Information is the lifeblood of any society." It is then highly problematic if this "lifeblood" is intentionally or unintentionally manipulated. The audience does not know what to believe, and the trust in sources of information is violated. Information disorders can be further classified according to intention, veracity, and potential harm (Egelhofer & Lecheler, 2019; Levak, 2020). The strategies for spreading these fake messages vary and include, for

K. Turková (✉) • V. Macková
Department of Journalism, Faculty of Social Sciences, Institute of Communication Studies and Journalism, Charles University, Prague, Czech Republic
e-mail: katerina.turkova@fsv.cuni.cz; veronika.mackova@fsv.cuni.cz

M. Wenzel, K. Stasiuk-Krajewska (eds.), *Disinformation in Central Europe*, https://doi.org/10.1007/978-3-031-76871-2_4

example, leaking, lying, seeding, and smearing (Arce, 2024; Boyd-Barrett, 2019).

The aforementioned problem has been accelerated by the rise of online communication. Online space and social media serve as effective tools for the distribution of disinformation and enable the delivery of false messages to wide audiences and various target groups faster than ever before. Additionally, social media contributes to grouping people with similar values and opinions, who then believe the information gained from community opinion leaders without proper verification (Aïmeur et al., 2023; Bennett & Livingston, 2020). The line between freedom of speech and preventing social and economic damage is thin, and regulation in this area can be understood as limiting human rights, especially in former communist states with censorship experience.

This chapter aims to describe the spread of propaganda, fake news, and disinformation in Central Europe. The goal is also to reveal the strategies used to disseminate these messages in the region, amongst societies that have been vulnerable to disinformation over the last two decades (Bokša, 2019; Levak, 2020; Rechtik & Mareš, 2021). Finally, the strategies for facing these threats that have been applied in Central European countries are summarized. Although research focusing on strategies of disinformation in the area exists (e.g., Bánkuty-Balogh, 2021; Syrovátka, 2021; Gregor & Mlejnková, 2021c; Wenzel et al., 2024), the current state and development of this phenomenon is still understudied, especially in the light of the problems that the region has to face.

The chapter is divided into four subchapters. First, the spread of disinformation in the region is examined, taking into account the historical context, and the important milestones related to the topic are highlighted. The two subsequent parts are dedicated to overviews of the disinformation narratives that accompanied the COVID-19 pandemic and war in Ukraine in Central European countries. To end on a positive note, the last subchapter provides a list of actions and initiatives that have been taken to fight against disinformation in this region.

4.2 The Spread of Disinformation in Central Europe

Central Europe began to be intensely targeted by information warfare from Russia after the Russian-Georgian war in 2008. The situation in the region was further impacted by Russia's illegal annexation of Crimea in 2014 (Bokša, 2019; Rechtik & Mareš, 2021). Due to the lack of clarity and the immediacy of the case, propaganda, as well as misinformation and disinformation narratives, circulated in society and were shared online, especially on social media. Troll armies, also called Kremlin trolls, were used by the Russian side to manipulate public opinion through online propaganda (Pavlíková et al., 2021). Despite the fact that mainstream and reputable media were able to publish the correct and verified information after a short period of uncertainty, the alternative scene was already established and the aforementioned narratives began to flourish (Gregor & Mlejnková, 2021a, 2021c).

Central European countries have been vulnerable to fake news and propaganda from Russia for various reasons, including their communist past, media pluralism and weakness of civil societies. The narratives of Russian disinformation warfare are built on the ideas of Slavic unity, Ostalgia (nostalgia connected with the Soviet Union and the period of time before 1989) and the impossibility of finding the objective truth. Moreover, they also contain strong counter narratives against the European Union (EU), North Atlantic Treaty Organization (NATO) or the United States (Bokša, 2019).

The rise in the use of social media and respective mobile applications has played an important contribution to the expansion of disinformation narratives of all kinds. Additionally, having a wide range of sources and channels has made the recognition of truth complicated and even impossible. The so-called post-truth era caused, amongst others, by the algorithmic power of social media along with the ability to express any information or opinion to the public, has become relevant and problematic in the Central European context (Gregor & Mlejnková, 2021b; Moravec, 2020; Pavlíková et al., 2021). This fragmentation of the public has opened up space for alternative and disinformation media channels, providing alternative versions of truth built upon fake news or disinformation (Gregor & Mlejnková, 2021c). Furthermore, China has influenced the region through their companies and social media applications imported to Europe (Győri et al., 2020).

During the first phase of the COVID-19 pandemic, which lasted approximately from March to June 2020, conspiracy theories spread in Central Europe thanks to alternative media producers, especially those acting online. Those actors presented COVID-19 as a human-made weapon while blaming various sources and authorities, for example, Bill Gates, China, the United States or the EU. On the other hand, the skeptical voices argued that COVID-19 was not harmful at all and recommended not wearing masks and avoiding vaccination (Győri et al., 2020). This process, characterized by the escalating dissemination of misleading information, has also been called an infodemic, and continued during the following phases of the COVID-19 pandemic (Gregor & Mlejnková, 2021b). The researchers argue that this issue has had a similarly significant and tragic effect as the pandemic itself (Rosińska, 2021).

After the Russian invasion of Ukraine in February 2022, a new wave of fake news and propaganda started to penetrate Central Europe. The disinformation referred, amongst others, to the Ukrainian refugees who had to leave their homeland due to the war (Stępniewski, 2022; Wenzel et al., 2024). This movement recalled memories of the medialized 2015-16 migration crisis, although the support for Ukrainians was higher than the willingness to accept previously incoming Afghan or Somali refugees (Moise et al., 2024). The false claims that circulated in the Central European region contained, for example, the information that the Ukrainian authorities are fascist or Nazi, or that many refugees from Ukraine have a criminal past (Wenzel et al., 2024).

The current strategies for spreading disinformation in Central Europe recall and modify the narratives already used in the past; hence they consist of undermining trust in the mainstream, reputable media, and the democratic system in general. The defamation of official and established sources of information creates space for alternative interpretations, propaganda and disinformation. The radicalization and polarization of the public sphere allow the accentuation of the messages and their intermediate sharing in "befriended" communities (Bánkuty-Balogh, 2021; Mareš & Mlejnková, 2021).

4.3 Disinformation Related to COVID-19 in Central Europe

According to the citizen's movement called Čeští elfové [Czech Elves] that aims to reveal disinformation campaigns and Russian propaganda in Czech cyberspace, the first disinformation about COVID-19 appeared in Central Europe relatively early. From January 2020, conspiracy theories circulated in Russian cyberspace, accusing the West of spreading the disease in the world (Čeští elfové [Czech Elves], 2020). False messages often included support or criticism of institutions and other states. Specifically, the EU and the United States were frequently criticized, and China and Russia were supported (Syrovátka et al., 2020).

The infodemic then began to thrive in Central Europe. Excessive amounts of disinformation or misinformation about COVID-19 were spread in both traditional and online media, as well as on social media (World Health Organization, 2020). "The following have contributed to the flood of inaccurate or false information about the COVID-19 pandemic: (a) the global scope of the health issue, (b) the unknown origin and nature of the coronavirus, (c) the unpreparedness of public health authorities, including the WHO, to face a global pandemic, (d) the information overload of contemporary society, (e) the behavior of digital networking media producers, (f) the desire of some social actors to gain media or public attention, etc." (Infomore.cz, 2020).

Various actors have sought to use the pandemic for their own benefit. The Internet, social media, and other communication channels, along with easily accessible information, have greatly contributed to the spread of fake news and have become a psychological burden for individuals. Disinformation had the power to negatively affect people's health, such as false information about vaccinations, COVID-19 medication, or masks (Győri et al., 2020; Ministerstvo vnitra České republiky [Ministry of the Interior of the Czech Republic], 2024). These phenomena led to several serious problems, including a decrease of trust in medical authorities and experts, resulting in low vaccination rates in some Central European countries, for example, Poland (Oleksy et al., 2022). Many people rejected COVID-19 vaccines based on unsubstantiated claims, leading to further spread of the virus and worsening the pandemic situation not only in the region.

According to the Central European Digital Media Observatory (CEDMO) project report, based on data gained in June 2022, more than half of respondents in the Czech Republic supported vaccination against COVID-19. However, it was also worrying that over a quarter of the sample was against COVID-19 vaccination. In Poland, support for vaccination was strongly influenced by age and education; hence older people and respondents with higher levels of education were more supportive regarding the COVID-19 vaccination. In Slovakia, there was distrust towards vaccination and other measures to prevent the spread of coronavirus, as almost half of respondents thought the Slovak government was using COVID-19 to its own advantage (Turková & Macková, 2023).

The research conducted by Chałubińska-Jentkiewicz et al. (2023) confirmed that COVID-19 was the subject of the majority of health disinformation that circulated in Polish society in 2021. The narratives, believed by more than one-fifth of the people under study, contained the claim that COVID-19 had evolved due to the 3G, 4G, or 5G sites, and that COVID-19 vaccines cause autism; hence the ratio of vaccinated inhabitants remained very low (Oleksy et al., 2022). This could have led to an even more destructive impact of the COVID-19 pandemic in Poland.

The situation in the Czech Republic and Slovakia was similar. The most mentioned conspiracy theory in the Czech media was also connected with the installation of equipment for wireless data transmission in the 5G network. Another claim that appeared in the media was the misinformation that American billionaire Bill Gates was responsible for the dissemination of COVID-19. The rapid spread of this fake narrative was also influenced by Czech President Miloš Zeman, who talked about the fake claim in a negative sense, but in so doing, prompted people to search for it (Győri et al., 2020; Infomore.cz, 2021).

It was evident that some creators of disinformation had simply replaced one topic with another. These people, who had previously spread fake news and conspiracy theories related to politics or health topics, began to focus their attention on the COVID-19 pandemic. This transition may have been motivated by the desire to maintain the attention of their followers and increase their influence on social media. In the past, the Czech media had already experienced a similar turn, when in 2015, the topic of migration began to dominate the public space, displacing the negative stereotyping of Roma people. Obviously, the actors who spread negative attitudes towards these issues remained active and flexibly changed the objects of their interest. Their main goal was to be in opposition and fight against some "enemy" (Němcová Tejkalová et al., 2021).

The improvement of visual elements and the presentation of information influenced the environment of disinformation websites and social media, not omitting Central Europe. Researchers from the Infomore.cz project, concerning the COVID-19 infodemic in the Czech Republic, analyzed two communication platforms spreading disinformation. Otevři svou mysl [Open your mind] and ProtiProud [Against the Current] spread false posts and articles that arouse feelings of fear and threat. The latter website and Facebook page are based on the conspiracy theories at the turn of the twenty-first century, which are incorporated in the visual design and language. Consequently, users are likely to recognize that this is not quality journalism. The edited photographs and images, including collages, give an amateurish impression. In contrast, the communication platform Otevři svou mysl [Open your mind] deals only with the topic of COVID-19 and looks very credible. For example, there are videos featuring researchers and "scientific" evidence; hence it can be very difficult to recognize that it is offers misleading or fake information (Němcová Tejkalová et al., 2021).

4.4 Disinformation in Central Europe Related to the War in Ukraine

As Wenzel et al. (2024, p. 193) argue: "The Russian war in Ukraine is not only being waged by military means; it is also an economic and information war." After the annexation of Crimea in 2014, alternative media became influential in Czech society, and this influence has prevailed to the present day. These websites and social media channels commonly promote fake news and narratives supporting Russia (Cabada, 2022). The persuasive audiovisual content that was previously spread by radio and television channels under the Kremlin's influence is nowadays easily published and shared on various social media, such as YouTube or Telegram (Orttung & Nelson, 2019; Solopova et al., 2023). Even some narratives have prevailed from the past, in particular, the comparison of the war operation to genocide, or the claim that members of the Ukrainian army are Nazis (Kovacheva, 2023).

In the weeks prior and immediately after the Russian invasion of Ukraine in June 2022, the societies in Central European countries became polarized, which created favorable conditions for the dissemination of propaganda and fake news. This phenomenon was particularly noticeable in Slovakia, which borders Ukraine. Slovak opposition Members of Parliament, who published nationalistic, conservative, anti-liberal attitudes, communicated pro-Russian disinformation and conspiracy theories on their public Facebook pages. For example, they stated that Ukraine had provoked Russia and called out the war, that the United States had caused the Russian invasion of Ukraine, and that Slovakia w in bondage to foreign forces (e.g., the United States or NATO) (Lintner et al., 2023).

In the Czech Republic, the pro-Kremlin disinformation outlets took advantage of the country's dependence on energy and natural gas from the Russian Federation. The war in Ukraine led to an increase in energy prices, which caused citizens' concerns about further price escalation and depletion of resources (Paličková & Černoch, 2024). Regardless of the shutdown of numerous disinformation websites by the CZ.NIC domain administrator due to their potential threat to national security (CZ.NIC, 2022), these media were still able to communicate with their followers on social media and networking sites. The arguments promoted by the pro-Russian channels accented that sanctions imposed on Russia by EU countries would harm Europeans, and attempts to reduce dependence on Russian energy and gas would cause an unprecedented price increase (Paličková & Černoch, 2024).

Polish alternative media "prepared the ground" for future actions approximately two months before the invasion, by informing people about fictitious provocations by Ukraine authorities towards Russia. After the beginning of the war, narratives similar to those in the Czech Republic relating to the lack of energy and natural gas, were accented by the disinformation scene. Later on, narratives focusing on the immigration of Ukrainian refugees that could destroy the Polish labor and housing market, cultural heritage and tradition of the country in general, were circulated on a massive scale (Łukasik-Turecka, 2023; Zilinsky et al., 2024). The hashtag #StopUkrainisationPoland was frequently shared on social media (Łukasik-Turecka, 2023).

The quantitative comparative research conducted in June 2022 in three Central European countries—the Czech Republic, Slovakia, and Poland—confirmed that although the narratives that circulated in the area were often similar, the likelihood of their acceptance and further spread differed due to country-specific experiences. The Slovaks blamed the Russian Federation for the conflict much less than Czechs and Poles, while more than one-third of respondents claimed that Russia bore no responsibility for the conflict in Ukraine, and only another weaker third blamed Russia for the conflict. Slovaks also accused NATO or Ukraine. On the contrary, three-quarters of the Poles under study had no doubt about Russia's guilt. In the Czech Republic, the sample was divided in two, with the weaker half primarily blaming Russia for the war in Ukraine (Wenzel et al., 2024).

4.5 Fighting Disinformation in Central Europe

When it comes to fighting fake news and misleading information, the situation in Central Europe is somewhat paradoxical. Despite relatively high awareness of the issue, the potential impact is not taken equally seriously (Bokša, 2019). Still, due to the rise of disinformation both in terms of volume and efficiency, the Central European governments and official authorities have taken action. Institutional responses have taken various forms, but the most common are warnings or alerts.

The warnings and alerts issued in the Czech Republic have often been related to the hybrid threats coming from China or Russia. In 2019, the Director of the Czech Security Information Service (BIS), Michal Koudelka, accented that Russia might seek to influence that year's EU parliamentary elections. Two years later, in 2021, Czech Prime Minister Petr Fiala introduced the National strategy for dealing with hybrid action (including disinformation), which was subsequently adjusted to the current conditions and threats (Cabada, 2022; Ministerstvo obrany ČR [Ministry of Defence of the Czech Republic], 2021, 2023). The National Cyber and Information Security Agency (NÚKIB) issued a warning against the use of Huawei and ZTE the same year. These companies were found as potentially dangerous due to the probability of the spread of Chinese influence in the Czech Republic (Hanzelka & Pavlíková, 2021). For the same reason, NÚKIB recommends not using the Chinese social media application TikTok, especially on work devices, in 2023 (NÚKIB, 2023).

While the motivation of Czech officials is predominantly security-related, the Slovak government still wants to maintain some ties with Russia and other influential forces for economic reasons. The warnings and alerts in Slovakia are thus much less confrontational, and address solely threats to security, rather than the possibility of sharing disinformation and influence (Rechtik & Mareš, 2021). The Polish government issued the Cybersecurity Strategy of the Republic of Poland for 2019–2024, which should, amongst others, lead to a greater level of resilience of information systems in private and public sectors against threats from cyberspace (Chałubińska-Jentkiewicz et al., 2023; gov.pl, 2024).

Tackling disinformation is also part of the EU agenda (Levak, 2020). For example, the flagship project led by the East Stratcom Task Force called EUvsDisinfo has been identifying various disinformation within member states since 2015 (Bokša, 2019; EEAS, 2024). Besides this, the European Commission financially supports research projects focused on fighting fake news and disinformation, for instance, the Central European Digital Media Observatory (CEDMO), Resilient Media for Democracy in the Digital Age (ReMeD) or WeVerify (Chałubińska-Jentkiewicz et al., 2023; CEDMO, 2024; ReMed, 2024; WeVerify, 2024). A similar strategy is applied by NATO, which also actively counteracts hybrid threads (Rechtik & Mareš, 2021).

Besides official actions, civil society initiatives greatly contribute to the fight against disinformation. In the Czech Republic, we can name a platform called Demagog.cz, aimed at fact-checking political speeches. Similar initiatives also exist in Slovakia and Poland, where they operate under the domains Demagog.sk and Demagog.org respectively (Kuś & Barczyszyn-Madziarz, 2020; Syrovátka, 2021). Some civil society activities target specific demographic and age groups. For example, the Czech organization Fakescape created interactive games for primary and secondary school students that cover topics of media literacy, cybersecurity, and disinformation (Fakescape, 2024).

4.6 Discussion and Conclusion

It is evident that disinformation is one of the most widespread disorders of the current era, especially in Central Europe. The ecosystem allowing the spread of propaganda, fake news, and disinformation in the region was established due to external, especially pro-Kremlin pressure after the Russian-Georgian war in 2008, and was reinforced after the illegal annexation of Crimea by Russia a decade ago (Bokša, 2019). The two recent events presented in this chapter—the COVID-19 pandemic and the war in Ukraine that began in February 2022—provided opportunities for the disinformation scene in Central Europe to expand, allowing the rapid spread of fake narratives, especially via social media.

In the Central European region, the COVID-19 pandemic, as well as the disease itself, were weaponized for the dissemination of fake news and disinformation. These messages criticized the EU and the United States, while China and Russia were supported (Syrovátka et al., 2020). The commonly shared narratives in the Central European region were related to the COVID-19 medication or vaccination, and the possible risks to people's health (Győri et al., 2020; Ministerstvo vnitra České republiky [Ministry of the Interior of the Czech Republic], 2024). Apart from other disinformation shared by the inhabitants of the region, such as the claim about COVID-19 being spread by 5G networks, the aforementioned negatively influenced people's willingness to use the available medical solutions to prevent the disease and possibly worsened the effects of the pandemic (Oleksy et al., 2022).

The war in Ukraine, which began in June 2022, recalled the pro-Kremlin action on an alternative scene that had already been witnessed in Central Europe. The preparation of the ground started well before the Russian invasion, when narratives about Ukrainian provocation appeared in pro-Russian media and social media channels. Later on, disinformation aimed at instilling fear in society flourished, with some messages being country-specific. For example, in the Czech Republic, alternative media wrote about the upcoming massive price increase caused by the lack of energy and natural gas from Russia (Paličková & Černoch, 2024). In Poland, these channels intensely incited hatred towards Ukrainian refugees (Łukasik-Turecka, 2023; Zilinsky et al., 2024). In Slovakia, even Members of Parliament in opposition distributed the message that Slovakia was in bondage to the United States or NATO (Lintner et al., 2023).

Nowadays, it is very difficult to recognize disinformation websites and social media channels, because many of them look professional and trustworthy. Their strategy is to present fictitious information as "scientific" evidence, which can cause public harm. Additionally, the articles and posts are often well-written and use manipulative techniques such as half-truths, emotive language, misleading statistics, or "expert" insights (as in case of Otevři svou mysl [Open your mind]) to make them persuasive. Although these fake messages can easily be refuted by empirical evidence, online consumers typically trust the information and do not always verify it in other sources (Wenzel et al., 2024). The algorithmic power of social media also contributes to the spread of disinformation, for example, by favoring viral and controversial content. Moreover, people tend to seek out and share information confirming their opinions and beliefs. This behavior leads to closed information bubbles, in which fake news circulates more efficiently.

The fight against propaganda, fake news and disinformation in Central Europe is in progress. It includes official actions undertaken by national and international authorities, as well as initiatives of civil society. The level of action differs amongst Central European countries: the Slovak government tends to be slightly more tolerant towards pro-Kremlin campaigns than Czech and Polish officials, due to their desire to keep closer ties with Russia for economic reasons (Rechtik & Mareš, 2021). However, this fight for truthful "lifeblood" in the region should not stop. Similarly to freedom of speech, the right to obtain accurate and unmanipulated information is essential—particularly in former communist states with censorship experience.FundingThis chapter is part of an international project co-funded by the European Union (EU) (action no. 2020-EU-IA-0267). This chapter has received funding from the European Union's Horizon Europe Research and Innovation Programme under grant agreement No 101094742. Views and opinions expressed are, however, those of the authors only and do not necessarily reflect those of the EU. Neither the EU nor the granting authority can be held responsible for them.

References

Aïmeur, E., Amri, S., & Brassard, G. (2023). Fake news, disinformation and misinformation in social media: A review. *Social Network Analysis and Mining, 13*(1). https://doi.org/10.1007/s13278-023-01028-5

Arce, D. (2024). Disinformation strategies. *Defence and Peace Economics*, 1–14. https://doi.org/10.1080/10242694.2024.2302236

Bánkuty-Balogh, L. S. (2021). Novel technologies and geopolitical strategies: Disinformation narratives in the countries of the Visegrád group. *Politics in Central Europe, 17*(2), 165–195. https://doi.org/10.2478/pce-2021-0008

Bennett, W., & Livingston, S. (Eds.). (2020). *The disinformation age*. Cambridge University Press.

Bokša, M. (2019). *Russian information warfare in Central and Eastern Europe: Strategies, impact, countermeasures*. The German Marshall Fund of the United States. https://www.jstor.org/stable/pdf/resrep21238.pdf

Boyd-Barrett, O. (2019). Fake news and 'RussiaGate' discourses: Propaganda in the post-truth era. *Journalism, 20*(1), 87–91. https://doi.org/10.1177/1464884918806735

Cabada, L. (2022). Russian aggression against Ukraine as the accelerator in the systemic struggle against disinformation in Czechia. *Applied Cybersecurity & Internet Governance, 1*(1), 1–16. https://www.ceeol.com/search/article-detail?id=1167114

CEDMO. (2024). *Central European Digital Media Observatory (CEDMO)*. Retrieved May 27, 2024, from https://cedmohub.eu/

Čeští elfové [Czech Elves]. (2020). *Pravidelný přehled české dezinformační scény – leden 2020* [Regular overview of the Czech disinformation scene – January 2020]. Čeští elfové [Czech Elves]. Retrieved May 22, 2024, from https://cesti-elfove.cz/wp-content/uploads/Hlaseni-leden-2020.pdf

Chałubińska-Jentkiewicz, K., Soler, U., & Makuch, A. (2023). Disinformation in Polish society in 2021 – Trends, topics, channels of transmission. *Polish Political Science Yearbook, 52*(1), 93–107. https://doi.org/10.15804/ppsy202309

CZ.NIC. (2022). *Sdružení CZ.NIC zablokovalo weby šířící dezinformace spojené s rusko-ukrajinským konfliktem* [CZ.NIC association has blocked websites spreading disinformation related to the Russian-Ukrainian conflict]. CZ.NIC. Retrieved May 28, 2024, from https://www.nic.cz/page/4301/sdruzeni-cznic-zablokovalo-weby-sirici-dezinformace-spojene-s-rusko-ukrajinskym-konfliktem/

EEAS. (2024). *EUvsDisinfo*. Retrieved May 2, 2024, from https://euvsdisinfo.eu/

Egelhofer, J. L., & Lecheler, S. (2019). Fake news as a two-dimensional phenomenon: A framework and research agenda. *Annals of the International Communication Association, 43*(2), 97–116. https://doi.org/10.1080/23808985.2019.1602782

Fakescape. (2024). *Fakescape! Učíme mediální gramotnost a kyberbezpečnost hrou* [Fakecape! We learn media literacy and cybersecurity by play]. Retrieved May 30, 2024, from https://fakescape.cz/

gov.pl. (2024). *Cybersecurity strategy of the Republic of Poland for 2019–2024*. Ministry of Digital Affairs. Retrieved May 24, 2024, from https://www.gov.pl/attachment/6a4aafc6-e339-4cd5-a8e6-cd47257f02d8

Gregor, M., & Mlejnková, P. (Eds.). (2021a). *Challenging online propaganda and disinformation in the 21st century*. Palgrave Macmillan. https://doi.org/10.1007/978-3-030-58624-9

Gregor, M., & Mlejnková, P. (2021b). Explaining the challenge: From persuasion to relativisation. In *Challenging online propaganda and disinformation in the 21st century* (pp. 3–41). Palgrave Macmillan.

Gregor, M., & Mlejnková, P. (2021c). Facing disinformation: Narratives and manipulative techniques deployed in the Czech Republic. *Politics in Central Europe, 17*(3), 541–564. https://sciendo.com/article/10.2478/pce-2021-0023

Győri, L., Istrate, D., & Istrate, D. (2020). *Disinformation after the first wave of the coronavirus in Central Europe*. Political Capital. https://policycommons.net/artifacts/3679120/disinformation-after-the-first-wave-of-the-coronavirus-in-central-europe/4484962/

Hanzelka, J., & Pavlíková, M. (2021). Institutional responses of European countries. In *Challenging online propaganda and disinformation in the 21st century* (pp. 195–224). Palgrave Macmillan.

Infomore.cz. (2020). *Infodemie* [Infodemic]. Infomore.cz. Retrieved May 29, 2024, from https://www.infomore.cz/cs/slovnik/infodemie/

Infomore.cz. (2021). *Mýty spojené s COVID-19* [Myths connected to COVID-19]. Infomore.cz. Retrieved May 25, 2024, from https://www.infomore.cz/res/file/analyzy/20210308-myty-klasicka-media/20210308-myty-klasicka-media.pdf

Kovacheva, R. (2023). Disinformation on COVID-19 and the war in Ukraine: Shared narratives and channels of distribution. In *КОМУНИКАЦИЯТА И МЕДИИТЕ НА XXI ВЕК: ОБРАЗОВАТЕЛНИ И ПРОФЕСИОНАЛНИ ПРЕДИЗВИКАТЕЛСТВА* [Communication and media of the 21st century: Educational and professional challenges] (pp. 34–51). Faculty of Journalism and Mass Communication, Sofia University "St. Kliment Ohridski". https://research.uni-sofia.bg/xmlui/bitstream/handle/10506/2568/Sbornik%20Konferentsia%20FJMC_2023_v.3.pdf?sequence=43#page=34

Kuś, M., & Barczyszyn-Madziarz, P. (2020). Fact-checking initiatives as promoters of media and information literacy: The case of Poland. *Central European Journal of Communication, 26*, 249–265. https://www.ceeol.com/search/article-detail?id=876255

Levak, T. (2020). Disinformation in the new media system – Characteristics, forms, reasons for its dissemination and potential means of tackling the issue. *Medijska istraživanja [Media Studies], 26*(2), 29–58. https://doi.org/10.22572/mi.26.2.2

Lintner, T., Diviák, T., Nekardová, B., Lehotský, L., & Vašečka, M. (2023). Slovak MPs' response to the 2022 Russian invasion of Ukraine in light of conspiracy theories and the polarization of political discourse. *Humanities and Social Sciences Communications, 10*(1), 1–11. https://doi.org/10.1057/s41599-023-02276-8

Łukasik-Turecka, A. (2023). Disinformation in the Polish media space in the first year of Russia's full-scale aggression against Ukraine. *Przegląd Strategiczny [Strategic Review], 13*(16), 251–260.

Mareš, M., & Mlejnková, P. (2021). Propaganda and disinformation as a security threat. In *Challenging online propaganda and disinformation in the 21st century* (pp. 75–103). Palgrave Macmillan.

Ministerstvo obrany ČR [Ministry of Defence of the Czech Republic]. (2021). *Národní strategie pro čelení hybridnímu působení* [National strategy for dealing with hybrid action]. Ministerstvo obrany ČR [Ministry of Defence of the Czech Republic]. Retrieved May 27, 2024, from https://mocr.army.cz/images/id_40001_50000/46088/Bezpecnostni_strategie_Ceske_republiky_2023.pdf

Ministerstvo obrany ČR [Ministry of Defence of the Czech Republic]. (2023). *Bezpečnostní strategie České republiky* [Defence strategy of the Czech Republic]. Ministerstvo obrany ČR [Ministry of Defence of the Czech Republic]. Retrieved

May 27, 2024, from https://mocr.army.cz/assets/informacni-servis/zpravodajstvi/narodni-strategie-pro-celeni-hybridnimu-pusobeni.pdf

Ministerstvo vnitra České republiky [Ministry of the Interior of the Czech Republic]. (2024). *Koronavirus: Přehled hlavních dezinformačních sdělení* [Coronavirus: Overview of the main disinformation]. MV ČR. Retrieved May 22, 2024, from https://www.mvcr.cz/cthh/clanek/koronavirus-prehled-hlavnich-dezinformacnich-sdeleni.aspx

Moise, A. D., Dennison, J., & Kriesi, H. (2024). European attitudes to refugees after the Russian invasion of Ukraine. *West European Politics, 47*(2), 356–381. https://doi.org/10.1080/01402382.2023.2229688

Moravec, V. (2020). *Proměny novinářské etiky* [Changes in journalistic ethics]. Academia.

Němcová Tejkalová, A., Macková, V., & Nainová, V. (2021). *Na vlně strachu, negativity a dezinformací* [On a wave of fear, negativity and disinformation]. Infomore.cz. Retrieved May 25, 2024, from https://www.infomore.cz/res/file/analyzy/20210406-dezinformace/20210406-dezinformace.pdf

NÚKIB. (2023). *Aplikace TikTok představuje bezpečnostní hrozbu* [TikTok application is a security threat]. Národní úřad pro kybernetickou a informační bezpečnost [National Cyber and Information Security Agency]. Retrieved May 21, 2024, from https://nukib.gov.cz/cs/infoservis/hrozby/1941-aplikace-tiktok-predstavuje-bezpecnostni-hrozbu/

Oleksy, T., Wnuk, A., Gambin, M., Łyś, A., Bargiel-Matusiewicz, K., & Pisula, E. (2022). Barriers and facilitators of willingness to vaccinate against COVID-19: Role of prosociality, authoritarianism and conspiracy mentality. A four-wave longitudinal study. *Personality and Individual Differences, 190.* https://doi.org/10.1016/j.paid.2022.111524

Orttung, R. W., & Nelson, E. (2019). Russia Today's strategy and effectiveness on YouTube. *Post-Soviet Affairs, 35*(2), 77–92. https://doi.org/10.1080/1060586X.2018.1531650

Paličková, A., & Černoch, F. (2024). Gaslighting Europe: Russia's energy disinformation in the Czech Republic. *Energy Research & Social Science, 112.* https://doi.org/10.1016/j.erss.2024.103497

Pavlíková, M., Šenkýřová, B., & Drmola, J. (2021). Propaganda and disinformation go online. In *Challenging online propaganda and disinformation in the 21st century* (pp. 43–74). Palgrave Macmillan.

Rechtik, M., & Mareš, M. (2021). Russian disinformation threat: Comparative case study of Czech and Slovak approaches. *Journal of Comparative Politics, 14*(1), 4–19. https://is.muni.cz/publication/1716176/JCPRechtik-Mares.pdf

ReMeD. (2024). *Resilient media for democracy in the digital age.* Retrieved May 24, 2024, from https://resilientmedia.eu/

Rosińska, K. A. (2021). Disinformation in Poland: Thematic classification based on content analysis of fake news from 2019. *Cyberpsychology: Journal of Psychosocial Research on Cyberspace, 15*(4), Article 5. https://doi.org/10.5817/CP2021-4-5

Solopova, V., Popescu, O. I., Benzmüller, C., & Landgraf, T. (2023). Automated multilingual detection of pro-kremlin propaganda in newspapers and telegram posts. *Datenbank Spektrum, 23*(1), 5–14. https://doi.org/10.1007/s13222-023-00437-2

Stępniewski, T. (2022). The Russia-Ukraine War, NATO's Eastern Flank, and Ukrainian Refugees in Central Europe. *Studia Europejskie, 26*(2), 7–15. https://www.ceeol.com/search/article-detail?id=1062744

Syrovátka, J. (2021). Civil society initiatives tackling disinformation: Experiences of Central European countries. In *Challenging online propaganda and disinformation in the 21st century* (pp. 225–253). Palgrave Macmillan. https://doi.org/10.1007/978-3-030-58624-9_8

Syrovátka, J., Pinkas, Š., Novák, A., & Uhrová, D. (2020). *Infodemic: Coronavirus and conspiracies in the Czech Republic*. PSSI. Retrieved May 25, 2024, from https://www.pssi.cz/publications/11

Turková, K., & Macková, V. (2023). *The effects and structure of disinformation on the population in connection with the COVID-19 pandemic in the Czech Republic, Poland and Slovakia (comparative report)*. CEDMO. Retrieved May 25, 2024, from https://cedmohub.eu/wp-content/uploads/2024/03/D5.1-EXTRA_The-effects-and-structure-of-individual-forms-of-disinformation-on-the-population-in-connection-with-the-COVID-19-pandemic-in-CZE-POL-SVK_comparative_eng.pdf

Wenzel, M., Stasiuk-Krajewska, K., Macková, V., & Turková, K. (2024). The penetration of Russian disinformation related to the war in Ukraine: Evidence from Poland, the Czech Republic and Slovakia. *International Political Science Review, 45*(2), 192–208. https://doi.org/10.1177/01925121231205259

WeVerify. (2024). *WeVerify – Wider and enhanced verification for you*. Retrieved May 24, 2024, from https://weverify.eu/

World Health Organization. (2020). *Call for action: Managing the infodemic*. WHO. Retrieved May 29, 2024, from https://www.who.int/news/item/11-12-2020-call-for-action-managing-the-infodemic

Zilinsky, J., Theocharis, Y., Pradel, F., Tulin, M., de Vreese, C., Aalberg, T., et al. (2024). Justifying an invasion: When is disinformation successful? *Political Communication*, 1–22. https://doi.org/10.1080/10584609.2024.2352483

CHAPTER 5

Linguistic Structures of Disinformation Messages in the Form of Fake News in Poland, Czech Republic and Slovakia

Karina Stasiuk-Krajewska, Peter Krajčovič, Ján Višňovský, Jana Radošinská, Martin Solík, Lenka Labudová, Ľuboš Greguš, Klára Zubková, and Lucia Furtáková

5.1 Introduction: Concept and Theoretical Context of the Study

The starting point of the comparative research presented in this chapter is the definition of fake news as a message that, by misrepresenting facts, structurally mimics journalistic news reporting, i.e. news (Lazer et al., 2018). Since the main function of news is constructed as an informational function, the genre is attributed specific characteristics in the form of objectivity and factuality (Dai et al., 2018). This discursive (operating in

K. Stasiuk-Krajewska (✉)
Faculty od Law and Communication, SWPS University, Wrocław, Poland
e-mail: kstasiuk-krajewska@swps.edu.pl

M. Wenzel, K. Stasiuk-Krajewska (eds.), *Disinformation in Central Europe*, https://doi.org/10.1007/978-3-031-76871-2_5

the consciousness of broadcasters and audiences, discursive and socio-cultural practices) correlation between news as a genre and the communication of unbiased information about the world is further strengthened by considering news as a model, constitutive journalistic genre (see: Chap. 2 of this book), and therefore a genre ascribed to a professional group that is constructed by reference to values such as truth, objectivity or independence (Allan, 2010; Stasiuk-Krajewska, 2018).

Accordingly, it was considered as the basis for this text that the features of fake news as texts that show similarities to the genre features of news should at the same time be considered as those that legitimise fake news in the news function. There is no doubt that these features, stemming from the typical discursive practices of journalism (e.g. checking the source) and referring to socio-cultural practices relevant to the understanding of democratic societies (e.g. the function of the media as guardians of the 'fourth power'), perform exactly such a function in real news.

The question of the characteristic features of real news as a genre is complex and has been the subject of detailed analysis in the second chapter. From the point of view of the presented research project and its implementation, it is relevant that genre is defined as 'a patterning of communication created by a combination of the individual (cognitive), social, and technical forces implicit in a recurring communicative

P. Krajčovič • L. Labudová
Department of Marketing Communication, Faculty of Mass Media Communication, University of Ss. Cyril and Methodius in Trnava, Trnava, Slovakia
e-mail: peter.krajcovic@ucm.sk; lenka.labudova@ucm.sk

J. Višňovský • J. Radošinská • M. Solík • Ľ. Greguš
Department of Mass Media Communication, Faculty of Mass Media Communication, University of Ss. Cyril and Methodius in Trnava, Trnava, Slovakia
e-mail: jan.visnovsky@ucm.sk; jana.radosinska@ucm.sk; martin.solik@ucm.sk; lubos.gregus@ucm.sk

K. Zubková • L. Furtáková
Department of Media Education, Faculty of Mass Media Communication, University of Ss. Cyril and Methodius in Trnava, Trnava, Slovakia
e-mail: zubkova1@ucm.sk; furtakova1@ucm.sk

situation. A genre structures communication by creating shared expectations about the form and content of the interaction, thus easing the burden of production and interpretation' (Ihlström Eriksson, 2004: 6). Thus, in analysing genres we analyse the texts, but also the discursive practices (related to the production and reception of these texts) and the sociocultural practices that locate these texts in the dominant worldview of contemporary societies. Such an approach leads to a maximalist research programme, whereby, starting from the study of typical textual structures (of greater or lesser complexity—from lexis or grammar to narratives and discourses), it is necessary to move on to the study of the transmitting and receiving mechanisms (encoding and decoding) associated with the texts. The social (discursive) practices understood must in turn be embedded in the cognitive field, primarily cognitive models such as stereotypes or other patterns of thinking, for example concerning cause—effect relationships (see Chap. 2; van Dijk, 1985, 1988). Cognitive models, in turn, cannot be analysed in isolation from the dominant ways of interpreting the social world, hegemonic discourses, especially the ideologies that form the axiological background of discursive communities, including professional communities (Gans, 1979; Soloski, 1989). Such holistic and coherent analyses have been conducted successfully in relation to specific and relatively closed discursive communities and their communication mediated by specific genres. There is also a rich literature relating to journalism and the news genre in these contexts (for example: Deuze, 2005; Zelizer, 2004; Reese, 1990; Hall, 1982; Schudson, 1978; more: in Chap. 6) although it is less structured and more multifaceted, for obvious reasons of the greater complexity of the phenomenon (e.g. the difficulty of defining, also institutionally, the boundaries of the journalistic profession).

In the case of fake news, such a programme at its maximum would include research not only into the structure of texts, but also into the specificities of the production, distribution and reception of such messages. It would also be necessary to look at the correlation between these practices and the cognitive and social. For example, the correlation between the characteristics of individuals and their susceptibility to disinformation messages; between existing social stereotypes and disinformation content; the cognitive structures into which disinformation content fits (e.g. so-called conspiracy theories). The institutional contexts in which fake news as a genre locates itself would also be an issue.

However, the point of focus in this chapter is the analysis of the texts considered to be representative examples of the fake news genre in terms of their structure and semantics, superstructures and macrostructures (van Dijk 1983, 2019). Of course, as discussed extensively in the second chapter, contemporary linguistic and especially discursive textual analysis cannot detach a text from its social contexts, especially the functions it performs. This applies to every syntactic and semantic structure. This is why phenomena from the level of language (of varying degrees of complexity) are analysed here as functional, therefore in connection with extra-linguistic reality. This is because we assume that 'A 'pure' structural analysis is a rather irrelevant theoretical exercise as long as we cannot relate textual structures with those of the cognitive and socio-cultural contexts of news production and reception. The development of linguistics and discourse analysis in the 1970s has shown, indeed, that a 'context-free' approach to language, for instance in the construction of formal grammars, is one-sided at best and certainly empirically inadequate. Of course, the same holds for the analysis of news discourse. It is impossible to really account for the many specific constraints on news structures, without specifying their social (institutional, professional) conditions or their socio-cognitive functions in mass mediated communication' (van Dijk, 1985: 71).

One of the essential functions of these structures and semantics, especially in the context of informational (news) or quasi-informational (fake news) media genres, is the question of legitimising these genres as presenting verified knowledge about the world, relating to facts. Legitimisation, generally speaking, is the act of justifying or explaining a social practice or policy. However legitimisation is also important for discourses, understood both as texts (the aforementioned legitimisation of textual functions) and as constructions of the social world (legitimisation of the interpretation of reality, of discourse in the theoretical sense). The essence of legitimation is the answer to 'the spoken or unspoken questions 'Why should we do this?' or 'Why should we do this in this way?" (van Leeuwen, 2008a: 105). The basis of all legitimacy is the communicative behaviours, structures, and semantics that offer the receiver (but also the sender) the possibility to find answers to these questions.

In the field of discourse analysis, Theo van Leeuwen (2007, 2008a) proposed a model of legitimacy, outlining four main discursive (communicative) strategies for building legitimacy. This model includes the following strategies: authorisation, moral evaluation, rationalisation, and mythopoesis. Authorisation is realised by appealing to authority, personal or otherwise (a person in the function of an expert, law, tradition, custom). Moral evaluation is based on values and morality (thus relating the action of a given social actor to categories of right and wrong), and is realised by evaluation, abtraction or analogy, which construct moral grounds for the positive or negative evaluation of a given action or situation. Rationalisation legitimises social practice by reference to its rational ends and effects, data and evidence, and appeals to knowledge, claims and logical arguments. Mythopoesis is legitimation achieved through the telling of short stories, actualising recognisable narrative models that expose most often the theme of punishment and reward (Van Leeuwen, 2007). Of course, the strategies presented can equally well serve the purpose of delegitimisation—by deprecating authority figures, judging the behaviour in question as morally wrong or irrational, or, finally, telling stories in which the protagonist suffers punishment for inappropriate behaviour. The analysis of the legitimation strategies used in the text has been used in interesting ways, thus proving its usefulness, for example in the analysis of fake news concerning the 2019 Nigerian elections (Igwebuike & Chimuanya, 2020). This study showed the dominance of the first strategy described, the strategy of authorisation, thus the use of the authority and status of the person or institution that is treated as the source of the false information.

5.2 Research: Objectives, Corpus Structure and Methodology

In line with the theoretical framework and overall concept presented above, the aim of the research was to answer the overarching question of how fake news messages legitimise themselves as information messages (news). The research was conducted in three Central European countries (Poland, the Czech Republic and Slovakia), which made it possible, on the one hand, to obtain interesting data and conclusions of a comparative nature; and on the other, to formulate an answer to the question of whether the indicated structural and semantic features of fake news have, at least in the scope of the three analysed countries and languages, a transnational and translingual character.

The analysis presented here therefore encompasses both structural and semiotic elements, on the premise that 'Textual structure acts to implement the informational and interpersonal meanings carried by a text so as to achieve certain rhetorical or communicative objectives. Thus we are interested in how a given textual structure arranges and presents both informational and interpersonal meanings, in where given meanings are located in the movement from textual opening to closure, in how informational and interpersonal meanings interact and in whether sets of distinct meanings operate to establish stages of textual unfolding' (White, 1997: 111), because 'analysis of texts is concerned with both their meanings and their forms. Although it may be useful analytically to contrast these two aspects of texts, it is in reality difficult to separate them. Meanings are necessarily realised in forms, and differences in meaning entail differences in form. Conversely, it is a sensible working assumption' (Fairclough, 1995: 57).

However, in line with the assumptions presented in the previous chapter on discursive textual analysis, the starting point will be structural elements, which will then be linked to semiotic aspects.

In the ongoing study, two underlying contexts of structures and semantics are actualised: news genre and legitimacy. This means, firstly, that we look for structures and semantics that correlate (negatively or positively) with the structural, stylistic or discursive characteristics of real news as a genre. Secondly, we ask the question about their function in the context of the legitimation strategies described above. Taking into account the theories of the news genre present in the literature (see previous chapter), the aim related to the identification of legitimisation mechanisms as well as the limitations of the adopted research method, the following structural and semantic elements (features) of the message constructed as relevant and specific to the news genre were taken into account:

1. General structure
2. Headline and its syntactic and semiotic characteristics
3. Corpus of text: features indicative of a reporter's style or lack thereof (e.g. objectivity or emotionality)
4. Mechanisms for constructing legitimacy
5. Updated narratives / discourses (interpretive frameworks of reality)
6. Themes

The corpus consists of fake news from January 2023–April 2024,[1] which has been verified as such on the AFP fact-checking portal in Poland

[1] The time span of the corpus is related, on the one hand, to the completion date of the previous phase of the research (Stasiuk-Krajewska, 2024) and, on the other, to the planned

(https://sprawdzam.afp.com/), the Czech Republic (https://napravou-miru.afp.com/) and Slovakia (https://fakty.afp.com/). The corpus was built on the base of a total of 366 fake news items ($N = 366$). Formally, the analysis was performed with reference to the fact-checker archive, which is linked from publications posted on AFP websites). The corpus included 143 messages in Polish ($N_p = 143$), 92 messages in Czech ($N_{cz} = 92$) and 131 messages in Slovak ($N_s = 131$).

An important assumption in the creation of the corpus was that the materials analysed had previously been found to be untrue or misleading by professional fact-checkers. In this way, it was possible to build a corpus about which there is no doubt that it fulfils the requirements of belonging to the genre category of fake news (it therefore consists of texts which are wholly or partially misleading and, in terms of structure, mimic news, a footnote). Due to the impossibility of an unambiguous decision, the question of the sender's intention was ignored in the course of selecting the texts forming the corpus. Obviously, defining the constitutive features of the corpus in this way carries important limitations for the research, which sheds light only on a small part of disinformation practices (footnote). On the other hand, however, such a methodological decision avoids an overly broad and imprecise definition of the research field, which is an important problem in the context of scientific disinformation analysis (footnote).

In the context of the selection of research material, it is also important to note that the AFP is a global news agency that conducts fact-checking analyses according to consistent standards in all three countries studied, with the result that texts labelled as untrue or misleading are of the same nature and are selected according to the same criteria. Of course, the AFP does not check all potentially false news circulating in a given country, but it applies specific and consistent criteria for their selection. These criteria relate, on the one hand, to an assessment of the relevance of the disinformation message to the public and, on the other, to its reach and popularity

publication date of the research results. In terms of content, it was considered that such a time frame would be adequate to the issue and purpose of the research, due to the continuous nature of disinformation activities (including the production and verification of fake news). In view of this, it is difficult to distinguish a time frame whose boundaries could be justified by specific events occurring in the social or political world. On the other hand, structuring the corpus in this way made it possible to capture a range of phenomena that provoke disinformation activities, such as armed conflicts, elections or health issues, and to avoid inadequacies related to possible thematic dominants (such as in the COVID-19 period).

on the web. These criteria were also considered valuable in view of the objectives adopted by the research team, although of course such a decision led to the research results being dependent on the standards of work of the selected fact-checking organisation. It is worth adding, however, that the AFP, being a member of the International Fact-Checking Network (https://www.poynter.org/ifcn/), applies standards for the selection and analysis of disinformation content that are internationally recognised and consistent in the global fact-checking community.

In the context of the structure of the corpus, it is also worth noting that some of the messages analysed appeared in two or three language versions. It was considered that a separate language version makes it possible to analyse the message as a separate syntagms, all the more so as in many cases the forms of presentation were modified to a greater or lesser extent (e.g. the choice of graphic illustrations, the inclusion or not of some additional material or the grammatical or semantic structures of the text). In the analysed corpus, 31 messages were common to all three countries, more (44 materials) were common to Poland and Slovakia, while the largest number, 49 texts, were present in both the Slovak and the Czech corpus.

Two methodological approaches are most commonly used in the study of fake news, both of which yield important and interesting results, but also have their limitations. The first approach is based on automated language analysis, inspired by the CADS theory and methodology (Heppell et al., 2023; Mu et al., 2023). This method is mainly used to analyse issues related to the subject matter and style of fake news (Dai et al., 2018; Raza & Ding, 2022; Fruttaldo, 2017). This type of research also has an important applied aspect—the identified text features are then used in the design of automated tools for the detection of disinformation messages (for example: Rubin et al., 2015). It is worth noting, however, the fundamental difficulty of this type of research in the context of the presented data base, which is related to the relatively small volume of the analysed corpora. Also this type of research does not generally address the functional as well as the discursive aspects of the regularities indicated in the texts. Another difficulty in applying automatic text analysis methods in the context of the research presented here is the lack of comparative material in the form of real news corpus data in anlysed languages.

The second frequently used method is the case study (for example: Maci, 2019). Selected structural and semantic features of fake news are discussed by analysing specific, single examples or a few examples together, selected according to various criteria. This is also a fairly common approach in the analysis of the news genre, used for example by van Dijk, whose concept is an important inspiration for this text. This kind of approach certainly makes it possible to capture the textual aspect of genre (in the context of grammatical and semiotic structures) in relation to discursive and socio-cultural practices (discourses). A limitation of this type of analysis is, of course, its representativeness. From the point of view of the objectives of the present research, this limitation is very important.

In the context of the above set of concepts, the approach used in this text can be considered a mixture methodology. Compared to macro (CADS) and micro (case studies) approaches, my proposed methodology is located at the medium level of data description, This approach is evident in both data selection and analysis. The number of texts that formed the corpus fell between the standards of corpus linguistics and the detailed case studies approach, and the analysis approach combined semi-quantitative counting methods with an in-depth analysis of individual texts and text fragments to examine and contextualize the findings suggested by the frequency count. Davies (2014) takes a similar approach: 'This medial level of investigation has several advantages for research questions that are broad in scope. It allows the researcher to construct reasonably robust 'maps' of language patterns through the analysis of frequencies of language-feature occurrence, without the loss of rich contextual detail. The e approach is effective both to confirm that the patterns of language features proposed are indeed consistently present (the usual province of Corpus Linguistics) and to describe what these features 'look like' in the context of the news media genres in which they appear (pp. 80–81).

The aims of the presented research, i.e. the reconstruction of the fake news genre model (in the context of the news genre model) and the comparative analysis of the characteristic features of this type of messages in three countries (three languages), necessitated the use of a hybrid model, combining elements of qualitative and quantitative analysis. A corpus of over 350 texts was analysed, which made it impossible to use the case study method. On the other hand, the research objectives, the search for correlations between the structural and the semiotic, and the small size of the corpus limited the validity of using CADS methods.

Therefore, the decision was made to manually code the data. The coding procedure differed depending on the data being coded. When coding data on grammatical structures or the occurrence/non-occurrence of specific communicative phenomena (occurrence of title, number of sentences in the title, syntax of sentences, occurrence of punctuation marks, emoticons or capital letters), the coding was carried out by a single coder who had the appropriate degree in linguistics. The coding rules were described in detail and discussed and practised during the research team meetings.

In the case of semantic and discursive structures, the first step was to identify the relevant categories associated with this part of the study. In this case, it was not necessary to create a sample of the data set, as the adopted categories were based on the categories used in the first wave of research, covering fake news in Polish and Slovak between January 2020 and December 2022 (Stasiuk-Krajewska, 2024). The adequacy of the adopted analytical categories was therefore verified during the implementation of the research procedure of the first wave of research. Later in the research procedure the coding was conducted by two coders who were experienced in researching and coding fake news messages and attended a training meeting. The inter-coder reliability test archived a satisfying result of Cornbach's alpha >0.75.

5.3 Data Analysis Scheme

Taking into account the above theoretical assumptions, the problems and objectives of the research, as well as the methodological limitations, the following data analysis scheme has been prepared, which will be supplemented at this stage of the presentation with some necessary comments of a theoretical and methodological nature.

Superstructures

Stage 1: general structure
Stage 2: title (headline)
a/ title (headline)—syntax and semantics
b/ title (headline)—emotions
Stage 3: body of the text
b/ emotions (emoticons and capital letters)
a/ date (numbers)

Macrostrucutres

Stage 4: legitimisation (discursive frames)
Stage 5: discourses / narratives
Stage 6: themes

The structure of the analysis of visual form is presented below (Diagrams 5.1, 5.2, 5.3, 5.4, 5.5, 5.6, 5.7, 5.8, 5.9, and 5.10).

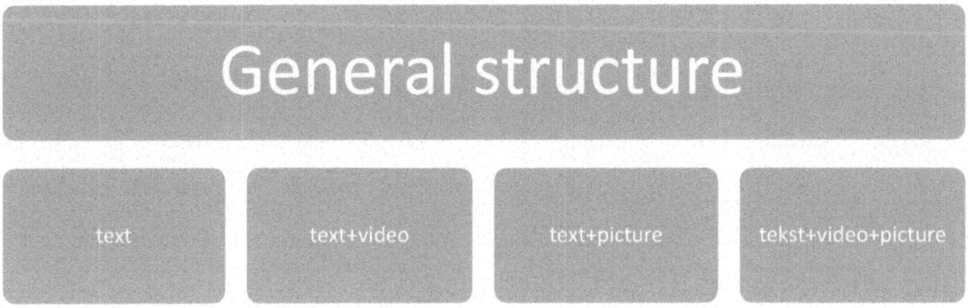

Diagram 5.1 General structure

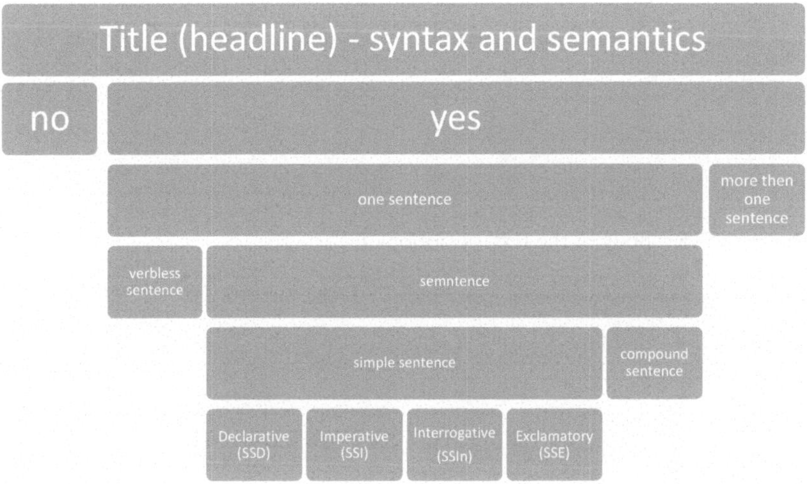

Diagram 5.2 Title (headline)—syntax and semantics

Diagram 5.3 Title (headline)—emotions 1

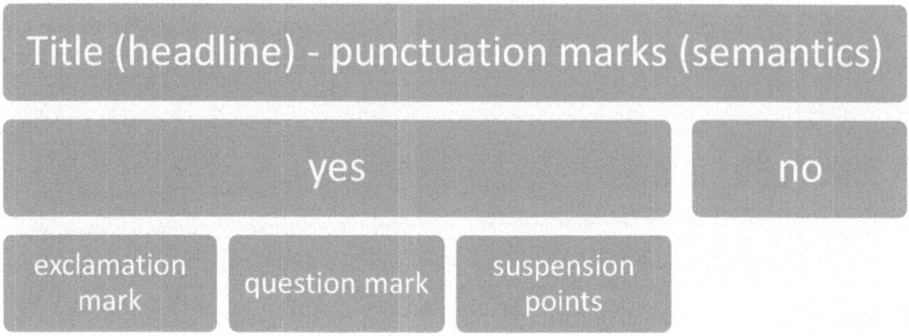

Diagram 5.4 Title (headline)—emotions 2

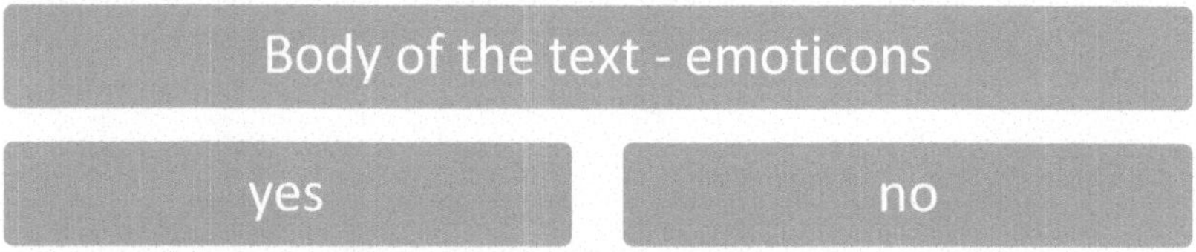

Diagram 5.5 Body of the text—emotions 1

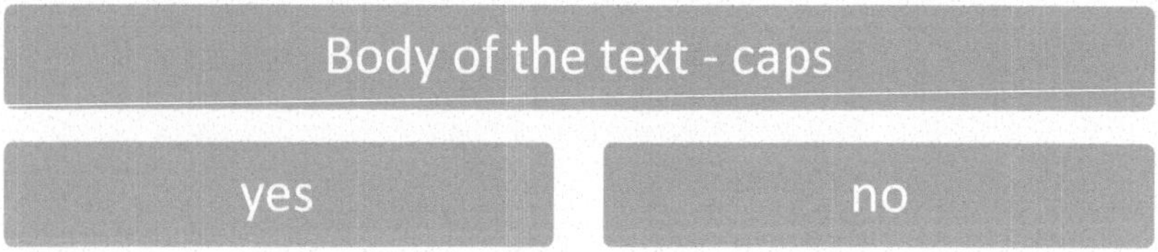

Diagram 5.6 Body of the text—emotions 2

Diagram 5.7 Body of the text—data (numbers)

Diagram 5.8 Legitimisation (discursive frames)

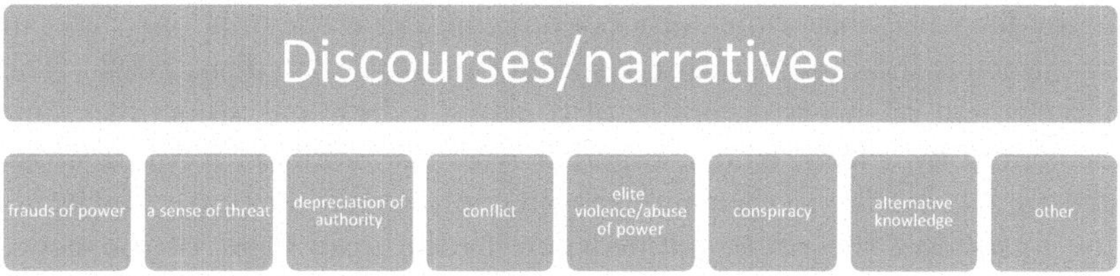

Diagram 5.9 Discourses/narratives

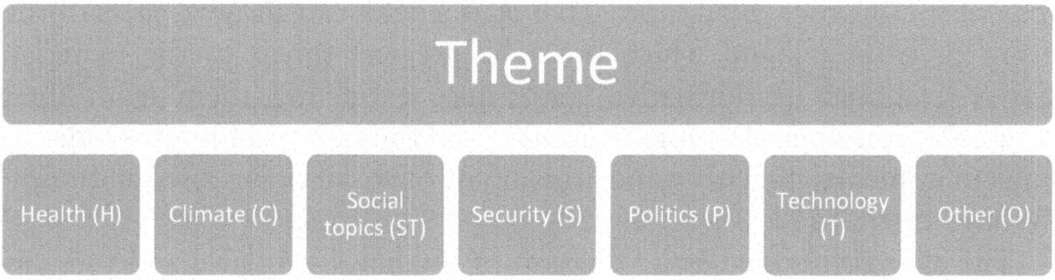

Diagram 5.10 Theme

The question of general structure is a question of the extent and degree of multimodality (about which more in the next chapter, footnote) of the texts analysed. This question is particularly relevant in the context of the widely recognised increase in the importance of visual and audiovisual elements in news media messages. There is also no doubt that such elements are very important in building the credibility of messages, becoming the essence of the so-called multimodal legitimisation, especially relevant in online journalism (Song & Lin, 2012). The issue of the specificity of the structure and function of visual messages will be analysed in the next chapter; here, let us just draw attention to the fact that the use of visual and audiovisual tools is becoming an increasing problem in the context of the recognition of disinformation messages, especially in connection with the development of artificial intelligence.

The extraction and more detailed analysis of the headline stems from the assumption that the structure of a hard news text can be divided into two main elements: the initial unit, containing the basic informational meanings, and the later stage, which is primarily aimed at qualifying, elaborating, clarifying and evaluating the meanings already presented in the initial piece, summary, and story (Yarlott et al., 2018: 27, Figure 1). This structure, reflected in the 'inverted pyramid' model, is so common to the genre described that some authors even consider it the only one that can be extracted and described as constitutive of news (Bateman, 2008). The first element of this structure is, of course, the headline, sometimes supplemented by a lead, which, together with the headline, forms a single structural unit (due to their similarity in content and function). The headline introduces the reader immediately to the main topic of the news, updates the discursive elements constitutive of news values (Bell, 1998; Molek-Kozakowska, 2013, 2014) 'goes directly to those aspects of the event or state-of-affairs which are assessed as constituting the peak or climax of social-order disruption. That is, it singles out those aspects of the event or issue at hand which pose the greatest threat to the material, power-relational or normative status-quo, extracting them from their original chronological or logical context and thus compelling the reader to engage immediately with some crisis point of social order disequilibrium' (White, 1997: 121).

The aforementioned basic division of the news structure and the particular role attributed to the headline justifies, firstly, the structure of the research conducted and, secondly, the particular importance given to the headline—since the initial part of the news is its most characteristic structural element: 'the news story has a beginning, but no end. Like the classic complication-resolution story, it begins with an abstract. But it does not end with a resolution and a coda. And the abstract, or 'lead' as it is called by journalists ('intro' in the US), does not serve to entice the listener or reader into the story, but instead conveys, without delay, the whole of the 'central event' of the story, as concisely as possible, in maximally 40, preferably only 30-odd words. Any 'orientation' is tightly wrapped into this, rather than presented separately' (van Leeuwen, 2008b: 352).

Elements of news articles are often referred to as moves. Moves are functional units that are used 'for some identifiable rhetorical purpose' (Connor & Mauranen, 1999: 51), so they contribute to the communicative (pragmatic) purpose of the text as a whole (Swales, 1990). If it is to be possible to distinguish them from other elements of the text, they should contain at least one proposition and some internal cohherence. Typical and constitutive of the news move is precisely the headline, which fulfils an informative purpose—it is supposed to inform the reader. It appears in all news, independent of the type of medium (Bednarek & Caple, 2012). However, its function is equally to attract and hold the reader's attention, since it is on the basis of the headline that the reader decides whether or not to read the rest of the text (Sala, 2023; van Dijk, 1988; Hartley, 2005; Isani, 2011).

The headline, which is the first 'point of contact' between the audience and the text, is also important in framing processes: it actualises a framework for interpreting the text in terms of cognitive processes and social knowledge more broadly (Claridge, 2010). As can be seen, the headline is functionally relevant at all three levels of discourse analysis: text, discursive practices (as evidenced, for example, by the fact that title writing is recognised as a particularly important journalistic competence) and sociocultural practices. For the reasons described above (the constitutiveness of the headline for the news genre, its important function in the construction of the relationship between text and audience, and finally, its significance in the context of broadly understood practices of interpreting the social world), it was decided to ask a question about the presence, specificity, and functions of the headline in fake news.

Connor and Mauranen (1999) describe the process of identifying moves as, firstly, identifying the rhetorical aims of the text (as a whole) and, secondly, dividing the text into meaningful units based on linguistic cues that determine the function and or boundary of the unit. While the task is relatively straightforward in the context of highly structured documents such as the EU grant applications analysed by the cited authors, it gets much more complicated in the case of news, especially fake news (for example: Kapusta et al., 2020). From a structural point of view, headlines must take into account the limited space allocated for their formulation. Consequently, to be attractive, they must be concise, formulated according to the rules of 'economy grammar' (Halliday, 2003). In addition to a number of syntactically detailed features, such as the frequent omission of prepositions, determiners, auxiliary verbs and the verb 'to be', the use of strong premodification, abbreviations, unconventional punctuation and a simplified tenses system (Sala, 2023), which are difficult to analyse on extended textual material (also because we are talking about relative phenomena, impossible to describe on a zero-one scale), the characteristic feature of headline syntax analysed in the presented research is the dominance of singular sentences over complex sentences and the occurrence of sentence equivalents.

Based on the assumptions presented above concerning the function and structure of the text, the following criteria were adopted for considering a piece of text as a headline:

- it is distinguished graphically from the rest of the text (message), most often by increased spacing between it and the rest of the text or by a series of emoticons;
- in the vast majority of cases, it is placed at the beginning of the text (it may or may not be placed underneath the visual or audiovisual material, but then it is the only textual element);
- it shall consist of no more than two sentences/sentence equivalents;
- the function of the headline—thus, in addition to attracting the viewer's attention, it also conveys the basic content of the material analysed, and thus has an informative function (for this reason, fragments such as: 'You haven't seen anything like this before', or 'What's going on here!' as the title of the film).

It is also worth noting that in the approach presented here, the text is seen as the result of a choice from a set of existing structural and semantic options. These are choices from among the available linguistic forms, from the lexical and grammatical potential (for example, choosing the passive form versus the active form). However, these formal decisions are at the same time choices of options from the potential for meaning (concerning how to present a particular event, how to relate to the potential audience of the text, what identities to project). Semantic choices are, in turn, linked to choices at the level of genre (which genres to draw on in creating a text) and discourses understood as interpellations of social reality (which interpretations refer to). Such a gaze also directs attention to what is absent in the text, to elements that have been omitted, those that have not been chosen (Fairclough, 1995). This is why both what is present in the text and what is missing from the text are so important. In designing the study, it was therefore considered that the presence of a characteristic news element such as the headline, in a similar form and function, in fake news would provide a bridge between these genres, thus legitimising fake news in its news function.

In the context of both of the structural element of fake news, the body of the text, the focus is primarily on encoding emotion and building the position of the sender as credible and informed.

The issue of emotion encoding is also of interest in the context of headlines because of its specific functions in relation to the sender. In the case of headline analysis, it is assumed that Speech function is related to grammatical mood, to the distinction between the main 'sentence types' (declarative, interrogative, imperative)' (Fairclough, 1995). The mood is encoded by the use of punctuation marks (or lack thereof). We realise, of course, that this relationship is not clear-cut, but in headlines, due to their structure and function, it seems to be relatively stable. This is due to the structural and functional peculiarities of the headline, which, being by definition a short form, at the same time loaded with semantics, strives for one-eyedness and building a direct (straightforward) relationship with the recipient as far as the sender's intentions are concerned. In the context of headlines, we can also see perfectly, the form of the sentence (declaratives, interrogatives, imperatives, etc.) builds up the semantics of the text—the relationship between sender and receiver, the frames.

The issue of the linguistic encoding of emotions is in itself very complex (Alba-Juez & Mackenzie, 2019), in the analysed corpus, in the case of the assessment of the emotionality of the body of the text (apart from the headline), the characteristics of the texts that could be unambiguously assessed were taken into account, such as the occurrence of emoticons (as an indicator of emotionality) or the use of capital letters (in a way that does not comply with the rules of orthography, as an expression of negative emotions).

In the search for textual determinants of the construction of the informational function of the text, the focus was on numerals, functioning in the analysed texts in the form of digits. This decision was based on the lack of doubt about the correlation that might occur here. It was also taken into account that the recommendation to provide numerical data is one of the important recommendations that concern the creation of a journalistic text that is valuable and interesting for the recipient (Bednarek & Caple, 2012).

A characterisation of the range of discursive frameworks that were identified as legitimacy-generating frameworks, together with their assignment to Van Leeuwen's legitimacy strategies (authorisation (A), moral evaluation (ME), rationalisation (R), and mythopoesis (M)) is presented in Table 5.1. As can be seen, given the specificity and complexity of the

Table 5.1 Legitimisation

Category	*Description*	*VN—model*
Evoking emotions	Positive or negative; most often fear, indignation, disgust, but also admiration or other positive emotions	ME
Building the position of the sender	The sender shares sensational news; shows off his or her knowledge by telling stories; shares life wisdom or an alternative interpretation of reality, the sender is clearly visible in the text, the essence of legitimacy is trust in him because of his unique knowledge of the world, life experience, moral qualities, etc.	Mix
Encouraging action	Open appeals or giving examples, including mass demonstrations	M
Appeal to authorities	Producing new authorities or heroes, in opposition to the 'official' ones	A
Presentation of evidence	Seemingly neutral, in the form of a photo, film, documents, scientific and other data, or eyewitness accounts	R

material analysed, it was decided to modify the proposed categories, the essence of which is to make them more specific in the context of the specific corpus of texts.

The characteristics of the discourses (the ways of constructing the social world to which the narrative refers or, in other words, the most important features, rules, values perceived in this world as regulating its functioning, on which the story is built) are presented in Table 5.2.

While superstructures or story patterns can be precisely described in terms of conventional communicative rules and strategies, when talking about legitimations, discourses and especially themes, we already move to the level of global units of meaning and therefore macrostructures. Macrostructure is the topic (or theme) of a globally, holistically understood text: 'The theoretical semantic macrostructure was introduced to capture that important aspect of discourse and discourse processing: It makes explicit the overall topics or themes of a text and at the same time defines what we could call the overall coherence of a text as well as its upshot or gist (Jufri & Usman, 2023: 13). In this context, of course, it should be noted that in many specific fake news stories, different thematic threads are interwoven. Thus, the theme, as understood by van Dijk and in this text, should be regarded as the thematic dominant, the answer to

Table 5.2 Discourses / narratives

Type	*Description*
Frauds of power	Presentation of the rigging of elections and other procedures
Alternative knowledge	Showing alternative knowledge about the world or alternative interpretations of events
Conspiracy	Uncovering conspiracies, speaking directly about the existence of a conspiracy (and not just fraud or alternate knowledge)
Elite violence/abuse of power/suppression of resistance	Showing restrictions and persecution
Conflict	Exposing social conflicts and hostilities—political, ethnic, class, etc.
Depreciation of authority	Stories about situations / statements / behaviour that bring public figures / institutions into disrepute
A sense of threat	Portraying something as a threat and creating a sense of anxiety or fear, destabilizing the sense of security

the common-sense question: what is the text about? Secondly, in general, the subject matter of news as a set of texts, or as a genre model, is diverse, though not coincidental—it stems firstly from the already described journalistic ideology and its associated prepositions, and secondly from a set of news values (Bell, 1991). News refers to what is constructed as socially relevant, but at the same time disrupts the natural, expected order of things. Accordingly, the broad dominant themes of hard news fall into three categories: aberrant damage, adversative rearrangements of power relations and normative breaches (White, 1997). The same theses can be successfully applied to the subject matter of fake news all the more so as they follow, relate thematically, and remain in relation to the dominant subject matter of real news at any given time. Put differently, disinformation reacts to the dominant media themes of the time, often raising the same issues, but presenting different interpretations and contexts.

The themes (macrostructures) identified in the study (based on the above theoretical assumptions and the analyses conducted in the first wave of the study, conducted only in Polish) are presented in Table 5.3. As can be seen, the structuring of the themes is not an unambiguous issue, which is why it was decided to make the categories more specific by creating subcategories ascribed to each thematic main category. The adopted

Table 5.3 Themes

Theme	*Examples*
Health	Covid-19, vaccinations, alternative therapies
Climate	Global warming; RES, green deal
Social topics	Immigrants, church/religion; social control (in contexts other than health); protests (in contexts other than health); LGBT+; economic issues; global conspiracy (in contexts other than health); ethnic minorities; nature (in contexts other than threats); public figures (non-politicians, members of governments; in contexts other than health)
Security	War in Ukraine, Israel-Hamas conflict, engagement of 'our' country in war, weaponry, the functioning of the army, Russian aggression in Ukraine
Politics	Politicians; political connections; elections
Technology	5G; AI and its interference with humanity, electric cars
Varia	Others, mixture, difficult to say

structuring is also in line with the type of classification of fake news topics adopted in studies that do not use automatic language analysis (Humprecht, 2019).

5.4 Research: Presentation and Analysis of Results

The presentation of the research results will be organised according to the data acquisition and analysis scheme. For the interpretation of these results, the explanations and clarifications made above are important and allow all the contexts, but also the difficulties of the interpretation process, to be fully taken into account.

Due to the fact that the aim of the research was both to obtain data concerning the structural and semantic elements of fake news as a genre and data which will make it possible to compare the structures and semantics of fake news occurring in Poland, the Czech Republic and Slovakia, the presentation of the research results will be in two stages. Firstly, data on the entire corpus of texts will be presented, and secondly, the data will be presented broken down by country.

5.4.1 Superstructures

5.4.1.1 Stage 1: General Format

In terms of general format analysis, the question asked is to what extent semantic elements in fake news that are distinct from the text are present,

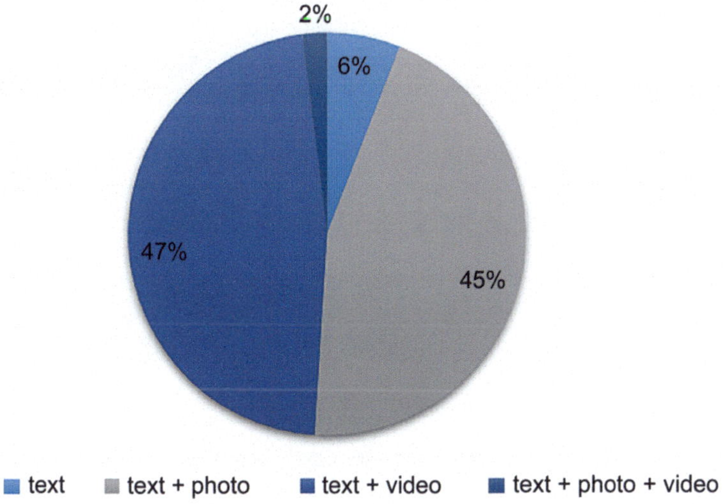

Chart 5.1 Format—entire corpus

and thus to what extent fake news is a multimodal genre. The multimodality of disinformation messages is recognised as an important feature of them, hence image analyses often accompany text analyses (Chart 5.1).

The structure of the analysed corpus in terms of overall format strongly confirms the thesis of the increasing role of visual messages in online communication. In the case of fake news, almost half of the messages are supplemented with a graphic element, what is also typical of news. Interestingly, however, slightly more (47%) fake news includes a video, which is not so typical of news. A negligible percentage of fake news includes both text and a photo or video, but interestingly a few more consist of text alone. The detailed functions and nature of the visual elements will be analysed in the next chapter, here let us just note that they occur both in the function of attracting the viewer's attention and increasing the attractiveness of the message and, very characteristically, in the function of lending credibility to the quasi-information contained in fake news. In addition, films in particular also evoke emotions. Thus, visual and audiovisual messages potentially have two different, if not contradictory, functions—emotional and informational. Therefore, a closer analysis of their occurrence in the messages analysed is necessary. It is also worth noting that visual messages are easier to receive, require less communicative competence, and are by

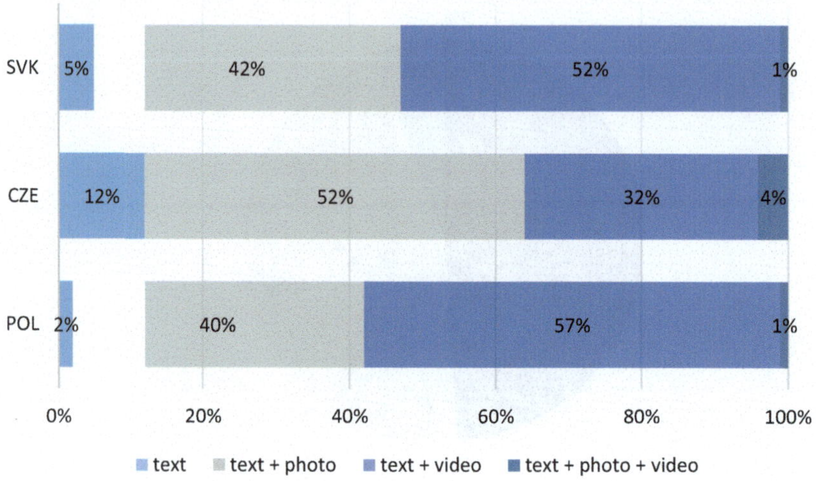

Chart 5.2 Format—comparison

their nature more persuasive. All this certainly provides justification for their prominent presence in fake news. Whether they bring the fake news genre closer to the news genre depends on their specific function, which will be analysed in the next chapter.

A comparison of fake news from the three countries studied reveals interesting differences (Chart 5.2). First of all, as can be seen, in the Czech Republic there are significantly more messages containing only text than in the other countries studied. This is primarily at the expense of messages that contain video, which in turn dominate in Poland. Interestingly, the number of messages containing photographs is also the highest in the Czech Republic. The structure revealed indicates that fake news in the Czech Republic most closely resembles news in terms of format, using text only to a relatively large extent and video to a relatively small extent.

5.4.1.2 Stage 2: Title (Headline)

In the context of the title (headline), the analyses were first concerned with whether the title (defined according to the criteria given earlier) appears at all in the analysed materials. Next, a question was asked about which sentence structures and which sentence forms predominate in the titles analysed. The question of encoding or not encoding emotions in them was also important. In this context, the number of sentences in the

Chart 5.3 Title (headline)—entire corpus

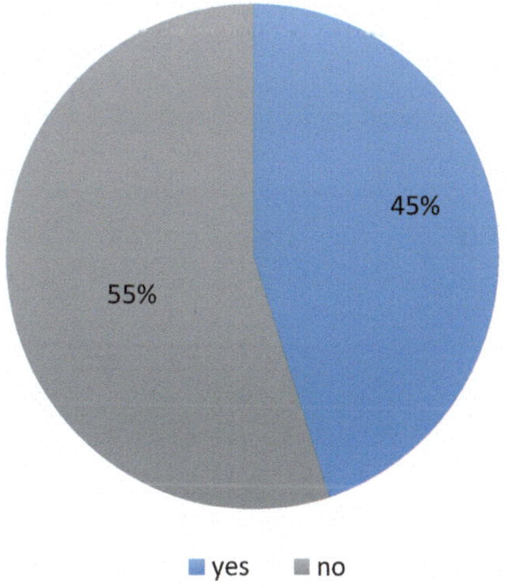

title, their syntax and their form were examined. The presence of emoticons in the titles was also analysed (Chart 5.3).

With a precise and restrictive definition of the title, it turns out that almost half of fake news messages have such an element, which both in terms of structure and function is identical (similar) to the headline in the news. Is this a lot? The answer to this question is not clear. However, let us remind ourselves that the headline is, both structurally and semantically, an inalienable element of news, so its absence might cause the recipients to feel a strong sense of the specificity and separateness of fake news in relation to news, while its presence strongly correlates with the information value of news. It is also worth noting that fake news is often relatively short, so it cannot be ruled out that the whole text has a structure similar to the headline and in fact functions as a headline—it attracts attention and summarises the basic information. The results certainly point to the need for further research into the semantics of titles (headlines) in fake news, especially since, according to data obtained using quantitative methods, it is different from that present in news (Horne & Adali, 2017) (Chart 5.4).

In a comparative context, one can again see the distinctiveness of fake news in Czech, where there are fewer titles than in Slovak or Polish. This is probably related to the lower presence of visual and audiovisual messages, which in the vast majority of cases are accompanied by a title. On

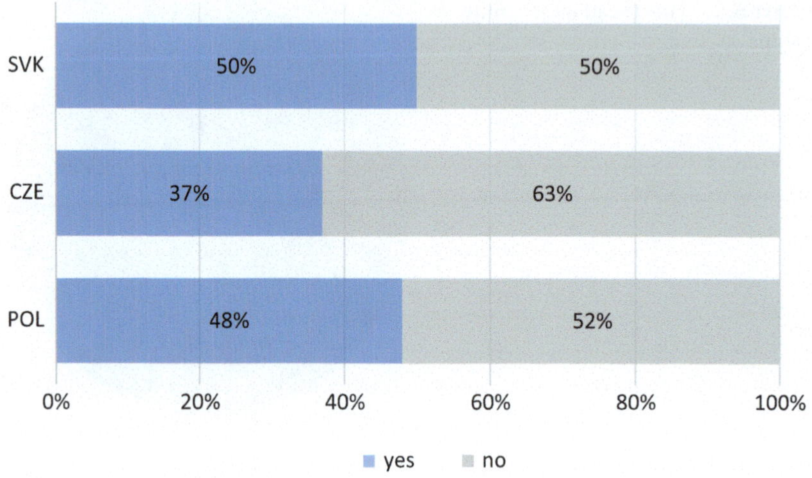

Chart 5.4 Title (headline)—comparison

the other hand, the strong presence of text and the relatively limited presence of video in Czech-language messages at the macro-structural level may compensate for the lack of a title when it comes to building the impression of informativeness in the recipient (Charts 5.5 and 5.6).

As can clearly be seen with regard to the first of the structural features of the title, the vast majority of the extracted text units consist of a single sentence. In this respect, no significant differences between the countries surveyed emerged either (the difference between Czech and Polish is 5%). The results obtained suggest that—if the title appears in fake news—its structural features (tendency towards abbreviation, ellipticality), associated with semantic and functional features (speed and ease of reception, brief presentation of the essence of the matter, etc.) are similar to the analogous features of the headline in news.

However, as can clearly be seen in Chart 5.7 almost half of the sentences that can be structurally and functionally distinguished as the title of fake news are compound sentences.[2] This is probably one of the communication system's suggested solutions to the problem faced by fake news

[2] Here we make the assumption that a simple sentence correlates with simple language, while a complex sentence correlates with language of a higher degree of complexity (plaintext).

Chart 5.5 Title (headline): sentences—entire corpus

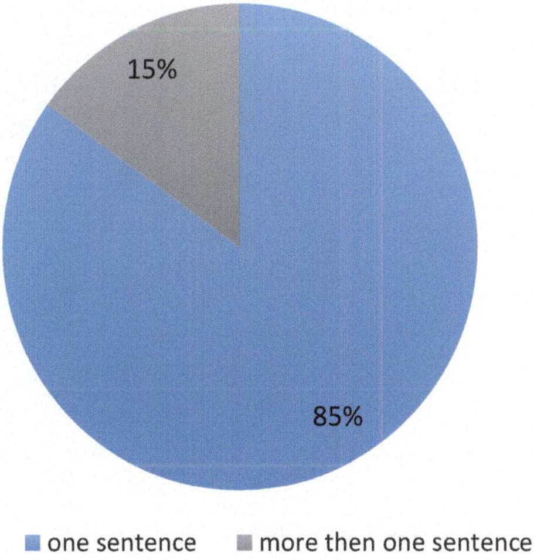

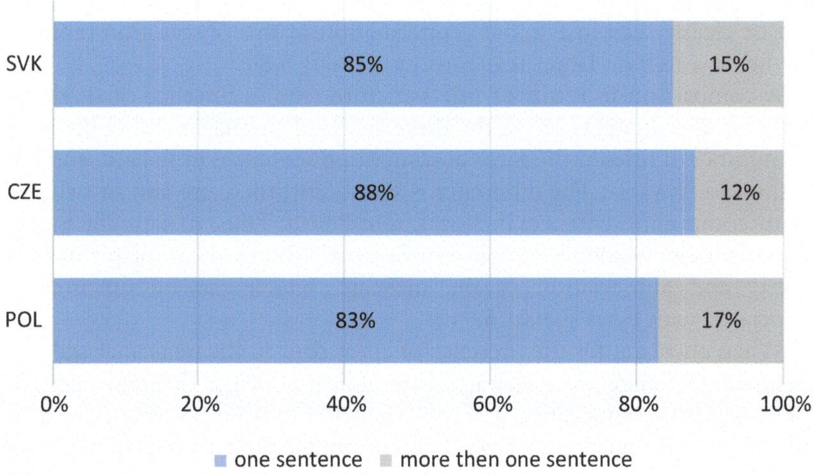

Chart 5.6 Title (headline): sentences—comparison

creators. If we create a title that consists of a single sentence, we are forced to expand this sentence to convey the basic information in it. This is a difficulty that real news creators also face. Let us note, however, that in online media this problem is most often solved in a different way—by staying with single sentences, but increasing their number to two in the title, as

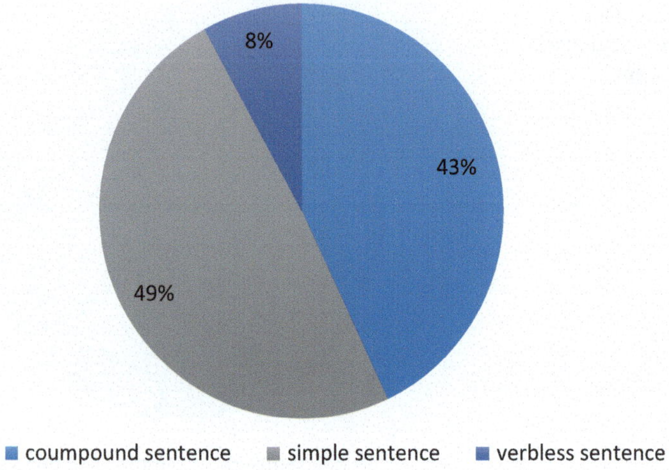

Chart 5.7 Title (headline): structure—entire corpus

can be clearly seen in Fig. 5.1 (confirmation of this observation requires further research on larger text corpora) (Chart 5.8).

A comparison in terms of title structure (the occurrence of single or compound sentences) across the three analysed languages shows the lowest number of titles in the form of compound sentences in Poland, and the highest in Slovakia. The difference is significant; there are almost twice as many compound sentences in Slovak as in Polish. Thus, structurally speaking, fake news headlines in Polish most closely follow the principle of simplicity and abbreviation of the message, which can be functionally associated with news (Chart 5.9).

When checking for the presence of emoticons in the texts analysed, we assumed that they are a characteristic indicator of the distinctiveness of fake news from news, since, as a rule, they do not occur in real news.[3] The appearance of emoticons, typical of communication mediated by social media, is most often associated with the encoding of emotions or a specific way of depicting them (through simplified pictograms, e.g. flags). Either way, emoticons are associated with strong signalling of the reception framework of a given text and as such are at odds with the language,

[3] The presence of emoticons was coded regardless of whether there was only one such character or more.

Fig. 5.1 Home page of one of the largest information portals in Poland, Onet. You can see the dominance of titles composed of two sentences. (Source: https://www.onet.pl/, 11.08.2024)

structure, and semantics of real news. Their absence in 2/3 of the titles analysed is a strong signal of the gravitation of fake news towards the news genre. It also indicates that fake news, at least at the level of the structural function of the title, does not directly encode emotions or evaluations. This conclusion is also supported by analyses of other short disinformation messages, such as tweets (Chart 5.10).

The comparative analysis shows that relatively more (although still far less than half) of all emoticons occur in fake news in Polish. Given the greater number of single sentences observed in this language in the titles, it can be assumed that emoticons supplement, in summary form, the

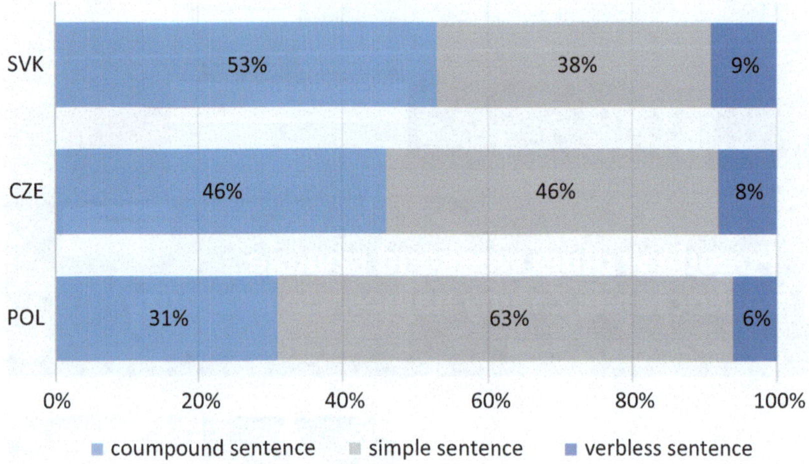

Chart 5.8 Title (headline): structure—comparison

Chart 5.9 Title (headline): emoticons—entire corpus

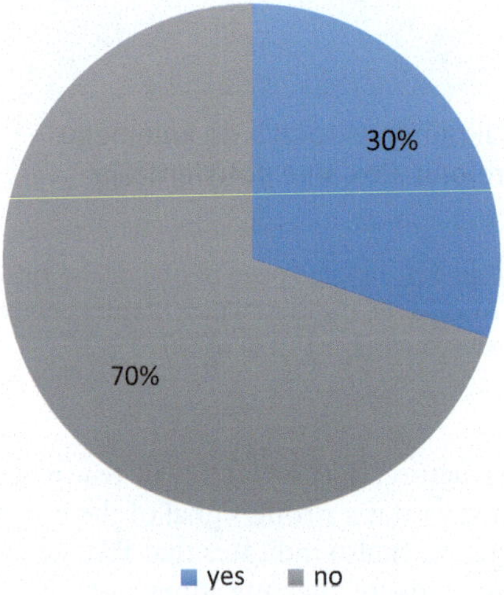

information that has not been conveyed linguistically (Charts 5.11 and 5.12).

Sentences in fake news titles are definitely dominated by those that do not end with any punctuation mark, or alternatively end with a full stop (this is most evident in Polish, and relatively least in Slovak). There are few exclamation marks and even fewer question marks. As mentioned earlier,

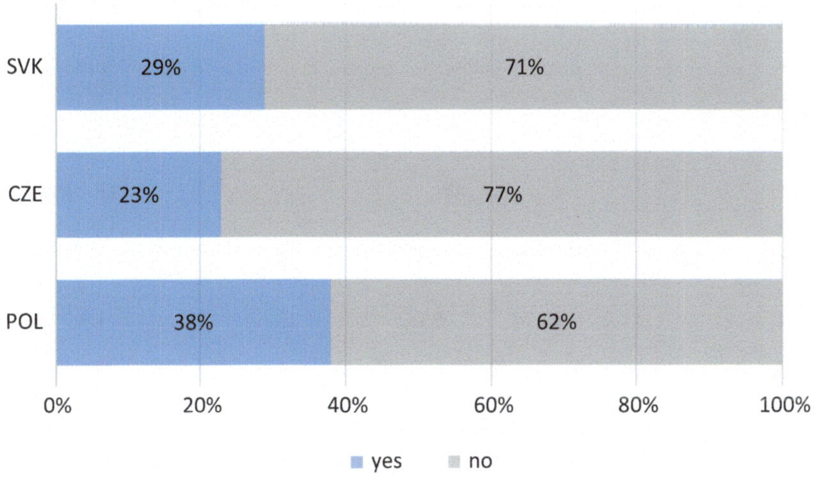

Chart 5.10 Title (headline): emoticons—comparison

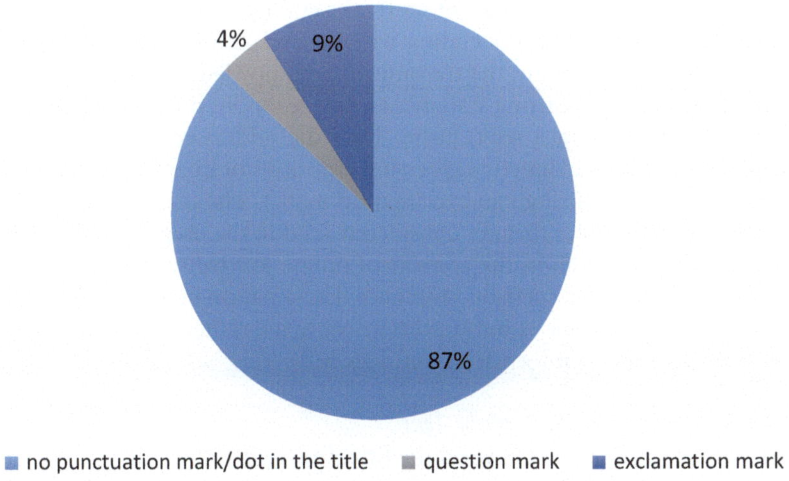

Chart 5.11 Title (headline): type from sentence—entire corpus

the occurrence of specific punctuation marks does not necessarily translate unequivocally into the communicative function of a sentence, but—generalising—sentences ending with a question mark, especially with an exclamation mark, encode modalities different from the one encoded by sentences without a punctuation mark or ending with a full stop. In this

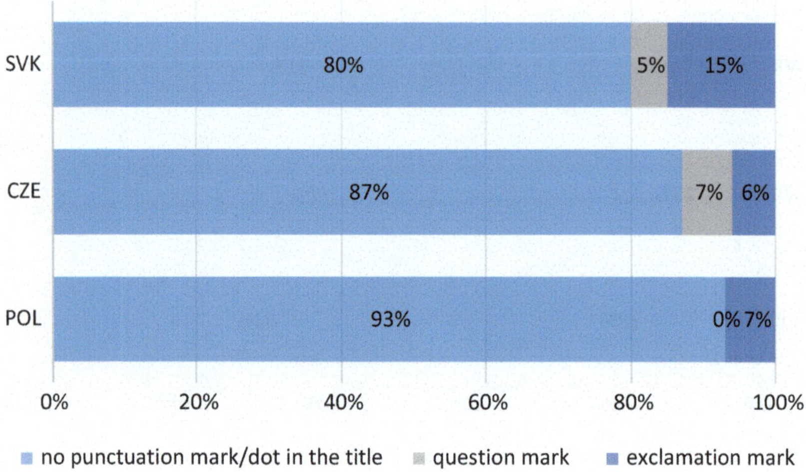

Chart 5.12 Title (type from sentence)—comparison

context, the strong dominance of sentences ending with the latter (against a small number of sentences ending with an exclamation mark or a question mark) suggests, firstly, the transmission of knowledge about the world or the formulation of strong statements, and secondly, it limits the linguistic means of building a relationship with the addressee (e.g. by asking him/her a question), directly expressing emotions or imposing something on the addressee (through an exclamation mark). The observed tendency is in line with the rules that are considered valid in the preparation of journalistic texts, although in the context of online journalism we observe an increasing number of question sentences. Exclamation marks are generally possible in very specific communication contexts, rather in the context of soft news and sports information (cf. Fig. 5.2)

5.4.1.3 Stage 3: The Body of the Text

As for the part of the text beyond the title, the body of the text (story), the emotionality encoded in it by means of structural and semantic elements was analysed in particular, taking into account the occurrence of emoticons and capital letters.

Secondly, the analysed presence of numerals in the texts, which was seen as a signal of the positioning of the author of the text as an informed person, and therefore as a way of building the credibility of the text and of realising its informative function (Charts 5.13 and 5.14).

Fig. 5.2 Sport news on one of the largest information portals in Poland, Onet. You can see the dominance of titles composed of two sentences. (Source: https://www.onet.pl/, 11.08.2024)

Chart 5.13 Emoticons—entire corpus

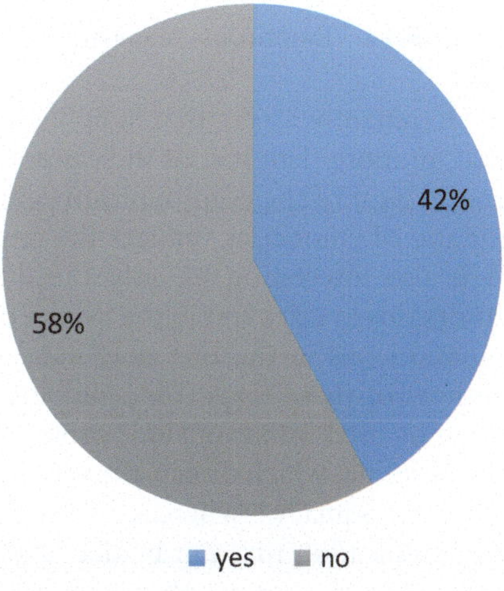

The number of emoticons in the entire text corpus does not show significant variation across the languages studied; in each case and in the entire corpus, it is higher than in the titles. It is worth noting, however, that emoticons are signs that occur in a minority of the analysed fake news texts (to the least extent in Polish, which confirms the thesis that in the titles they 'supplement' the semantic limitations of single sentences). This is interesting insofar as it undermines the belief that fake news is

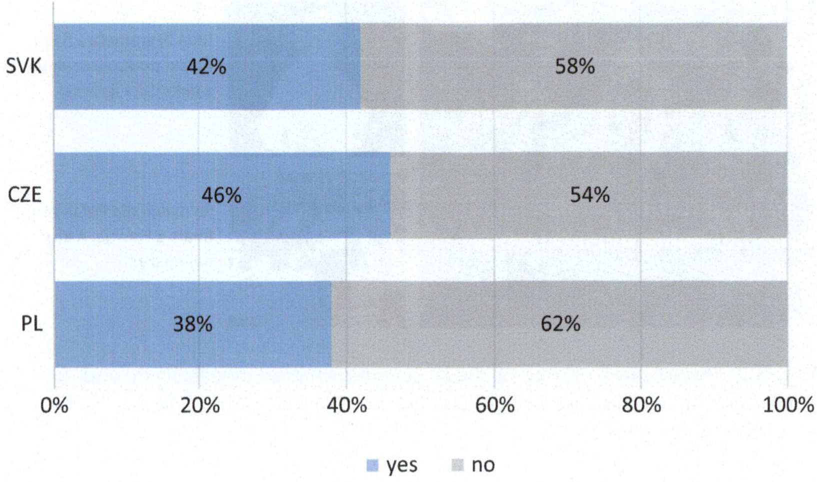

Chart 5.14 Emoticons—comparison

predominantly a text that simply encodes emotions. It can also be seen that, in more than half of the cases, they do not use signs that are considered typical of the communicative flow of social media. In both contexts, the use of emoticons reduces the credibility and 'seriousness' of the text. It is also interesting to note the difference between the occurrence of emoticons in titles and in the bodies of texts. The limitation of the number of emoticons in the title may indicate the discursive competence of the authors of these texts (this competence does not have to be realised by them or used intentionally), concerning the specific role and function of the headline, which directs the reception of the entire text. To put it somewhat colloquially—whereas in the headline, authors try to refrain from communication that can be decoded as 'frivolous', they allow themselves to do so in the corpus of texts (Chart 5.15).

To an even lesser extent, capital letters occur in the texts analysed (without justification in the orthographic rules). These were analysed on the assumption that—even more so than in the case of emoticons—capital letters encode (negative) emotions, a lack of objectivity and a lack of professionalism. They are also strongly associated with the language of social media. The use of capital letters very clearly positions the author as an emotionally involved, agitated or at least irritated person, seeking to share his or her emotions with the recipient. In view of the above, the absence of capital letters in 4/5 of the analysed fake news should be considered significant (Chart 5.16).

Chart 5.15 Caps—entire corpus

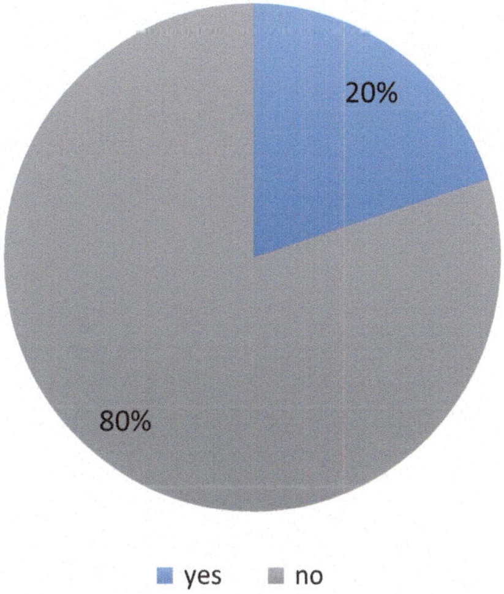

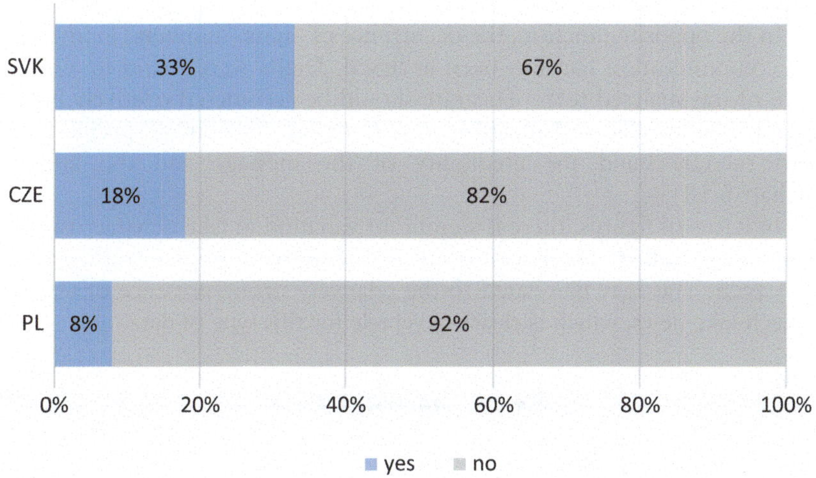

Chart 5.16 Caps—comparison

A comparative analysis, however, indicates that such average results may be misleading. Czech is closest to this average. There are by far the fewest capital letters in Polish and, surprisingly, a lot in Slovak. The difference is all the more significant because, as already mentioned, capital letters are a strong indicator of the emotionality of the text and the author's specific, involved, and non-objective attitude (Chart 5.17).

Chart 5.17 Numbers—entire corpus

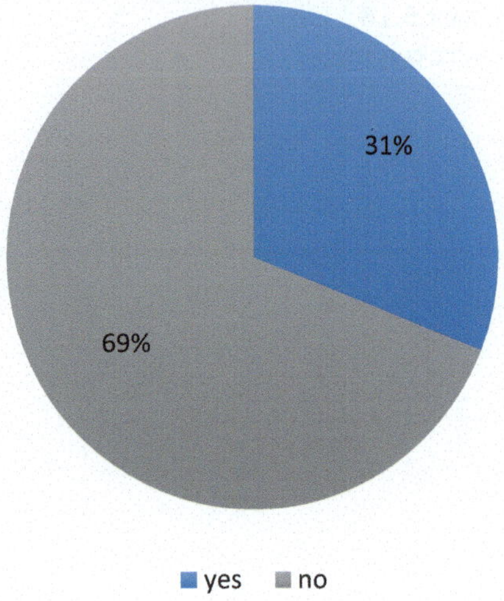

In the opposite function, the occurrence of digits (numbers) in linguistic communication has also been analysed. Digits were found in almost 1/3 of the analysed texts. This rate should be considered relatively high, given their specific and unambiguous function—to convey specific data and thereby build the credibility of the message and the author (Chart 5.18).

In terms of figures, there is significant variation in fake news across the languages studied. The fewest occur in Polish, with a significant increase in Czech. This may be related to the relatively strong presence of text in Czech fake news, which is clearly a vehicle for this type of data.

5.4.2 *Macrostructures*

5.4.2.1 Stage 4: Legitimisation (Discursive Frames)

The analysis of legitimacy takes place in relation to the general framework of the text and the framing process. Both the framing process (framing) and the framework itself refer to the socio-cultural aspects of discourse analysis at the most general level. This is why legitimacy analysis has been referred to as discourse framing analysis (Charts 5.19 and 5.20).

The methods of legitimisation used in fake news texts vary, although there is clearly a very significant predominance of legitimising the message

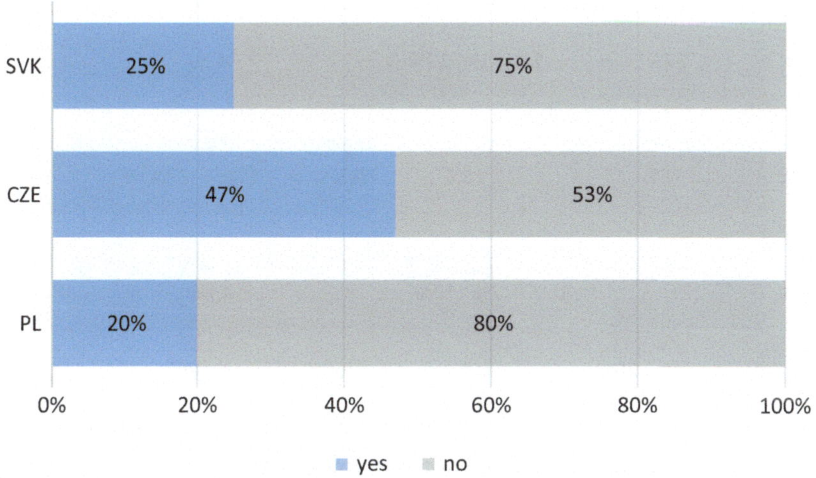

Chart 5.18 Numbers—comparison

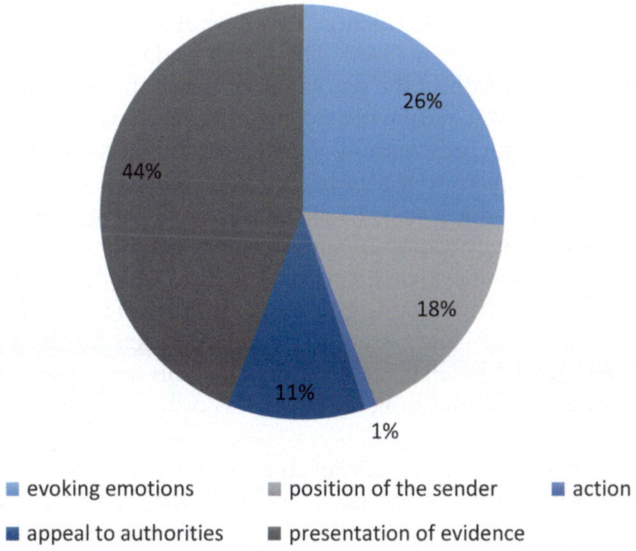

Chart 5.19 Legitimisation—entire corpus

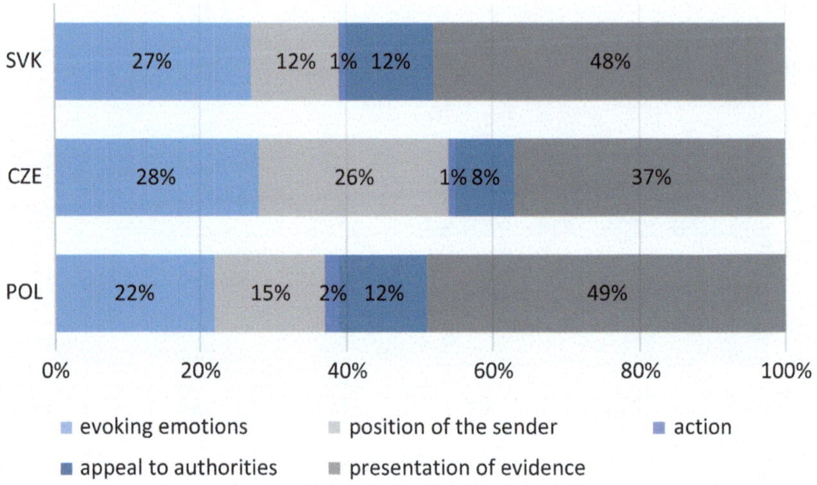

Chart 5.20 Legitimisation—comparison

through the presentation of evidence. This type of legitimisation accounts for almost half of the texts analysed. This is followed by evoking emotions and building the position of the sender. Note that both of these methods do not exclude the presentation of evidence. However, while in the case of legitimisation through presentation of evidence, special emphasis is placed on allowing the recipient to learn as much as possible about the (alleged) evidence (e.g. see a document, listen to someone's statement, watch a video), in the case of evocation of emotions, the main goal is precisely to evoke emotions, even if evidence is presented, little attention is paid to the accuracy of the presentation. Legitimising by referring to the author's position, on the other hand, assumes a strong position for the author in the text. The author can tell his or her story, share his or her life experience or, by means of communication, emphasise his or her superiority over the audience.

To reiterate, in the analysed corpus, the reference to legitimisation of a rational nature is relatively dominant. Legitimisation of this kind has another persuasive advantage—namely, it builds a communicative situation that leaves the recipient feeling that he or she has interpreted and evaluated the presented data independently. Characteristic is also the limited reference to authorities, which is probably due to the fact that one of the features of the world vision to which fake news refers is the negation of authorities, communicated as the so-called ability to think for oneself.

In comparative terms, no significant differences were revealed between the three countries, although fake news in Poland can be considered slightly more 'rational' than fake news in Slovakia and especially the Czech Republic.

The dominance of rational legitimacy definitely brings fake news closer to news in terms of macro-structures.

5.4.2.2 Stage 5: Discourses/Narratives

At this stage, the dominant visions of the world (discourses) to which fake news refers were analysed. These visions are operationalized in the form of narratives—hence the analysis was termed discourse/narrative analysis. Discourses and narratives refer to certain constitutive, in their internal perspective, features of the world. This world can thus be constructed as a dangerous place; a place ruled by conflict (at various levels) or the restriction of the individual either by false moral authorities or by direct abuses of power; and finally a place where true knowledge is concealed (often as part of some global conspiracy) and must be revealed without delay (Chart 5.21).

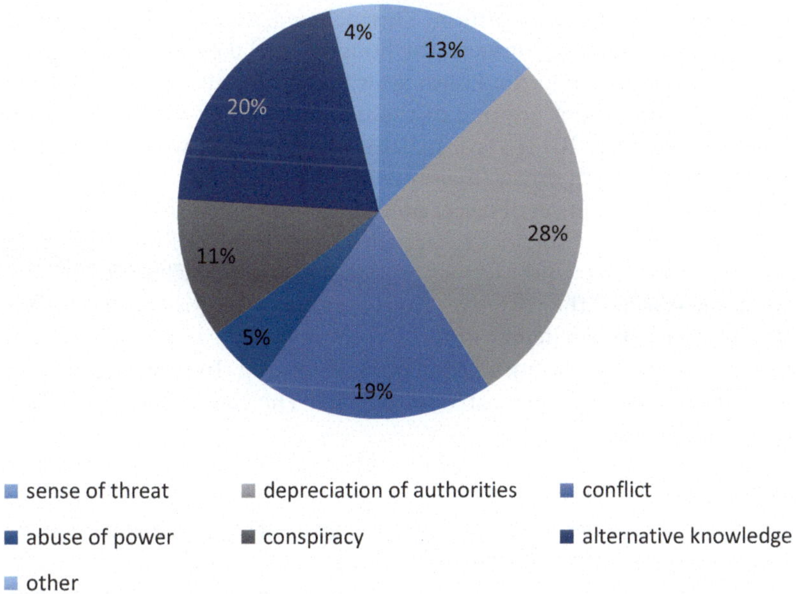

Chart 5.21 Discourses—entire corpus

The discursive (narrative) structure of fake news is varied. At the forefront, however, unsurprisingly in view of the observations made above regarding the denial of authority, is the narrative of the deprecation of authorities. The deprecated authorities are mainly politicians, however, those who either have a lot of power (for example, they are linked to the European Union) or are presented in the media discourse as unquestionable authorities. The most common theme in the material analysed was the deprecation of Volodymyr Zelensky and those associated with him (privately—his wife, for example, or politically—the Prosecutor General of Ukraine). To a very limited extent, the deprecation of authorities referred to local politicians or representatives of the scientific or medical community. Next in terms of frequency of occurrence in the analysed corpus were narratives presenting alternative knowledge of the world. It should be emphasised that only a minority of these narratives concerned alternative quasi-scientific knowledge (e.g. cancer treatment), the majority revealed, in the author's opinion, 'inconvenient' facts from broader social and political life—for example, child trafficking in Ukraine or planes getting stuck at the airport. Almost 20% of the narratives referred to the conflict. These were fake news that showed the world as a place of conflict between different social groups, while at the same time aiming to escalate this conflict, building an us vs. them opposition. This included, for example, the aggression of Ukrainians, alleged loans for housing for Ukrainians or the presence of LGBT people in the Polish army. Let's note that conflict and the creation of oppositions are one of the main structures actualized also in hard news (Cramer, 2011; Davies, 2014). This was followed by narratives that created a sense of threat (mainly concerning armed conflicts, but also, for example, the addition of insect products in the bread sold at Orlen stations) and conspiracy theories (especially in the context of the earthquake in Turkey, pandemics and vaccines or global warming) (Chart 5.22).

A comparison of the narratives from the three countries primarily indicates a lesser role for deprecation of authority in Poland, in contrast to Slovakia, where this narrative is exceptionally strong. In contrast, the conflict narrative is underrepresented in Slovakia. The Czech Republic is closest to the average.

5.4.2.3 *Stage 6: Themes*

The topic of fake news was analysed in the context of 6 main thematic categories (plus varia). Within each category, sub-categories were then distinguished, allowing the thematic scope of each main category to be presented accurately (Chart 5.23).

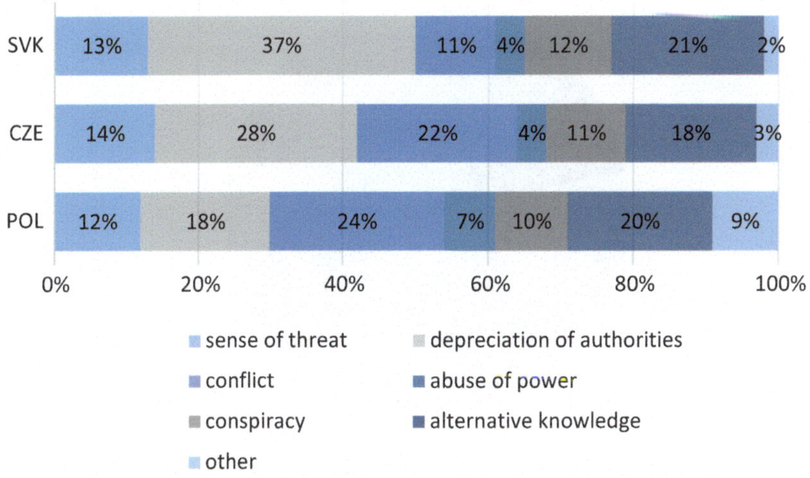

Chart 5.22 Discourses—comparison

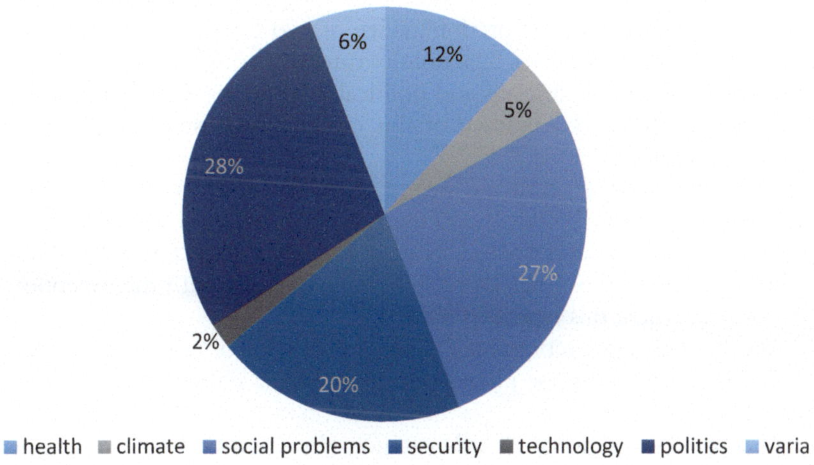

Chart 5.23 General themes—entire corpus

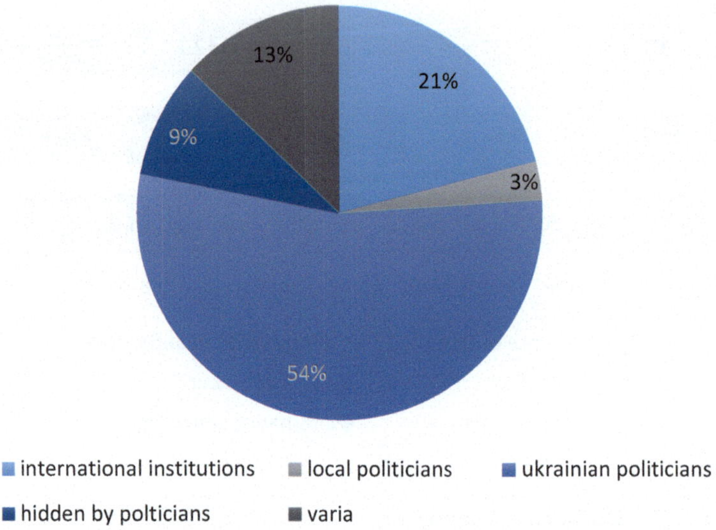

Chart 5.24 Politics Poland

In terms of themes, the predominance of two categories can be seen: politics and social problems. The politics category primarily includes messages about specific politicians and political institutions. On Charts 5.24, 5.25, and 5.26, the breakdown of this category into sub-categories by country is showed.

As can be clearly seen in the area of interest in politics, the subject of fake news is primarily Volodymyr Zelenski especially in the Czech Republic, but also in Slovakia and Poland. Noteworthy is the low interest in local politics and politicians (e.g. Donald Tusk in Poland), with the exception of Slovakia, where this type of topic reaches 31%.

Another strongly represented category is that of 'social problems' (27%). It refers to diverse phenomena that have a direct impact on citizens' lives.

A detailed breakdown of this category by country is shown in the Charts 5.27, 5.28, and 5.29.

In this case, the differences between fake news topics in the Czech Republic and fake news topics in Poland and Slovakia are clearly visible. The social themes are much less varied in the Czech Republic, with the messages focusing primarily (59%) on problems related to the everyday life

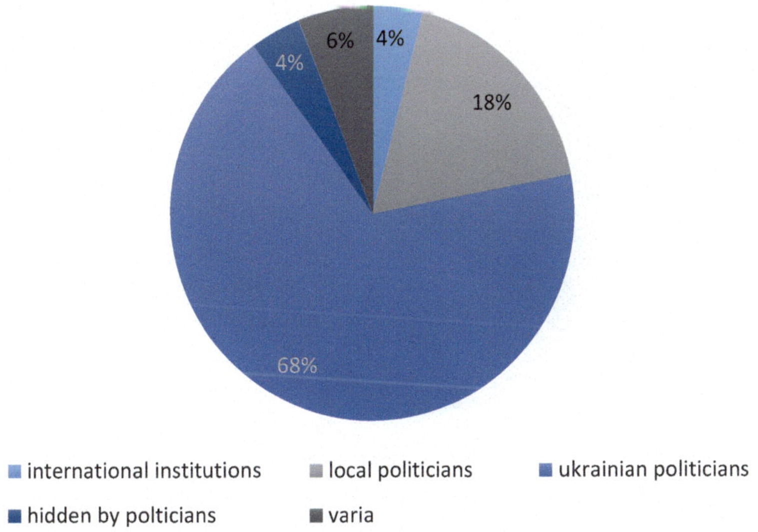

Chart 5.25 Politics Czech Republic

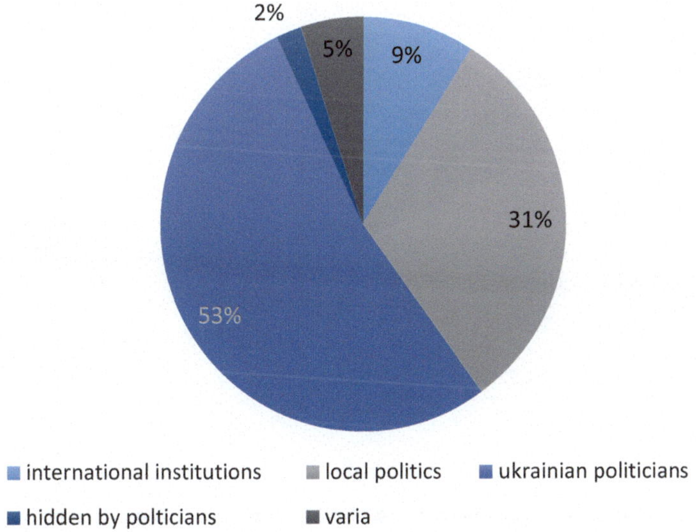

Chart 5.26 Politics Slovakia

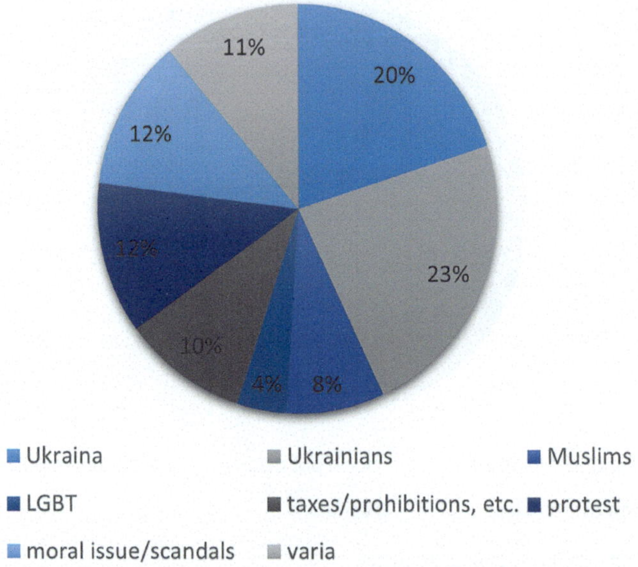

Chart 5.27 Social problems Poland

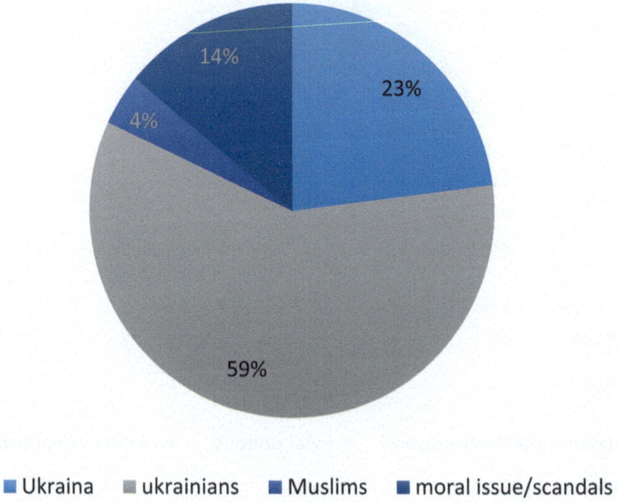

Chart 5.28 Social problems Czech Republic

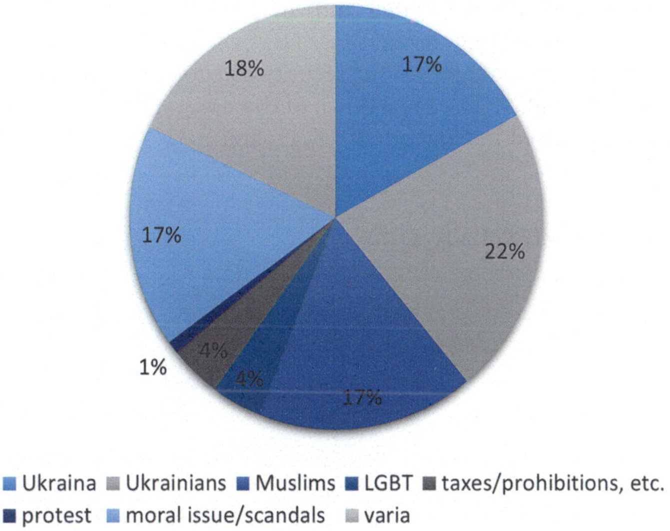

Chart 5.29 Social problems Slovakia

of refugees from Ukraine (the presence of Ukrainian children in schools, extra money for Ukrainians, problems with adaptation, cultural differences, etc.). These issues are also mentioned in fake news from Poland and Slovakia, but to a much lesser extent (23% and 22% respectively). In Czech fake news, there is also a strong presence of the theme of Ukraine, which is interpreted as difficulties or threats experienced by the citizens of a given country, but not related to the presence of refugees, but to state policy (e.g. transport of grain from Ukraine). In this case, however, the difference with Poland and Slovakia is not so high (23%, 30%, and 17% respectively). Moral scandals are also a theme present in Czech fake news (14%). This is a category, that refers to topics that arouse moral outrage or disgust (e.g. fake news about sex being another Olympic discipline approved by the IOC). In Poland, this category accounts for 12%, while in Slovakia, it accounts for 17%. A thematic subcategory related to Muslim refugees is also relatively strongly represented in all three countries. Other categories that are not present in Czech fake news discourse concern LGBT+ people, repression by political authorities and, to a small extent, rebellion and resistance movements (e.g. farmers).

Another thematic category is the security category, which deals primarily with armed conflict. The breakdown of this category into subcategories by country is showed in Charts 5.30, 5.31, and 5.32.

Chart 5.30 Security Poland

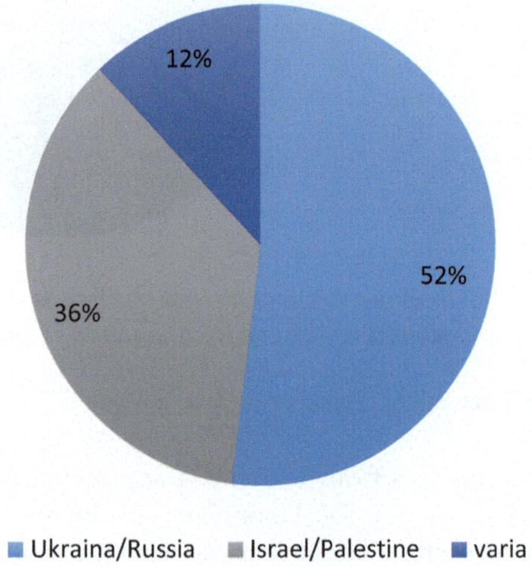

Chart 5.31 Security Czech Republic

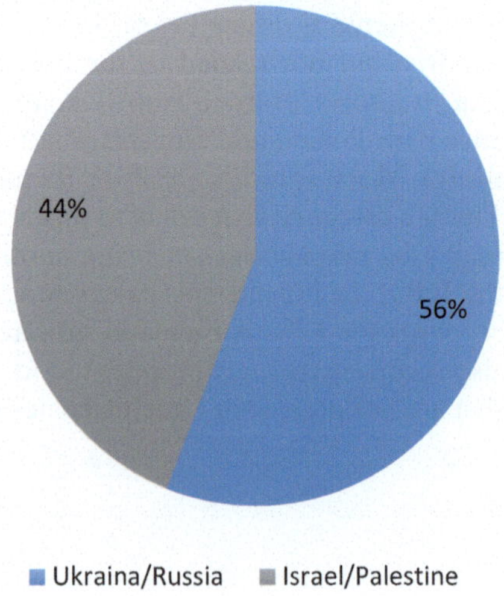

Chart 5.32 Security Slovakia

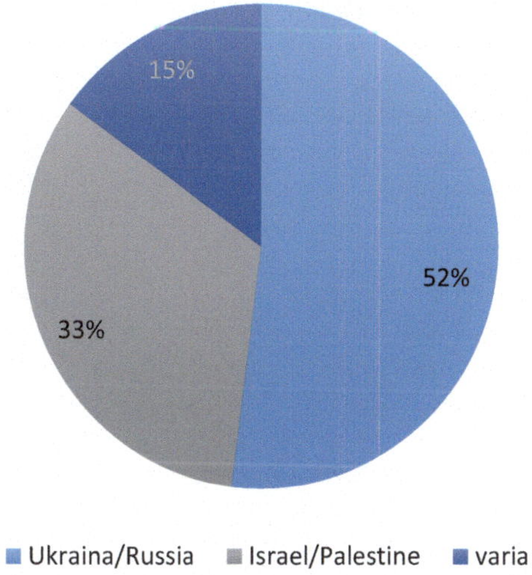

As can be seen, disinformation about Russia's aggression in Ukraine definitely prevails in this context, although the conflict between Israel and Palestine (Hamas) is also strongly present. In Slovakia and Poland, there are also other themes, related to natural disasters or mobilisation.

The last category to be analysed in detail by breaking it down into subcategories was health. Its presence throughout the corpus is not high, which is particularly symptomatic. Within this category, the category of vaccination was extracted (because, interestingly, Covid-19 in the corpus studied does not function as part of a global conspiracy or so-called 'plandemic', but through the prism of compulsory vaccination and its effects).

As can be seen (Charts 5.33 and 5.34), the vaccination theme is more strongly present in Slovakia and the Czech Republic than in Poland. What is particularly interesting, however, is the difference between the presence of the topic of Covid-19 and vaccination in the first wave of research and its presence today. Namely, we observe a radical decline in interest in this area. This is evidence that fake news is 'interested' in what is the subject of the most heated social debate and the cause of the strongest social controversies and fears. Fake news, in terms of subject matter, follows the news. In this sense, when we talk about this macro-structural element, fake news definitely copies the genre characteristics of news. This trend is, of course,

Chart 5.33 Health, Poland -

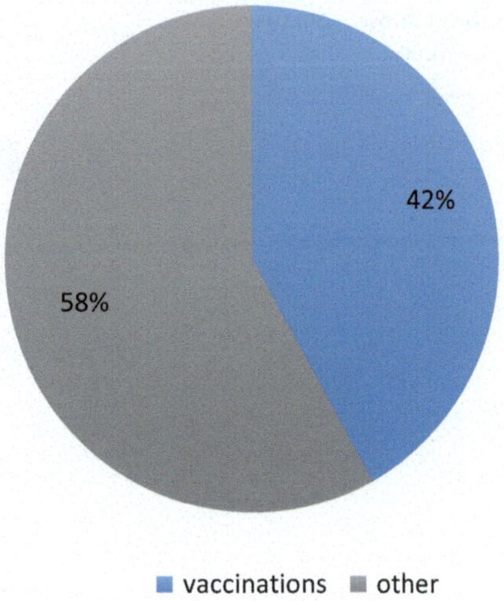

Chart 5.34 Health, Czech Republic

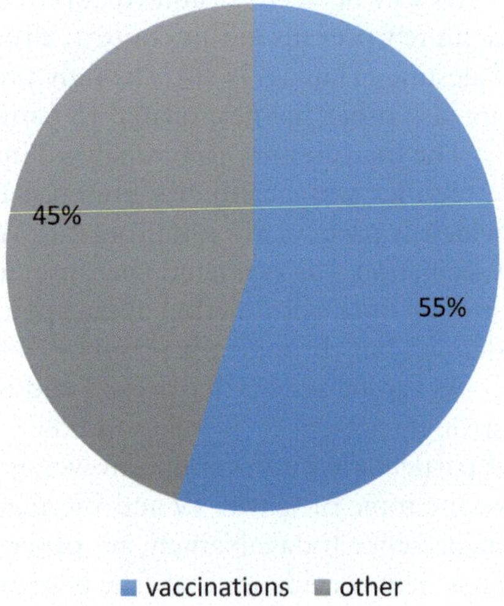

Chart 5.35 Health, Slovakia

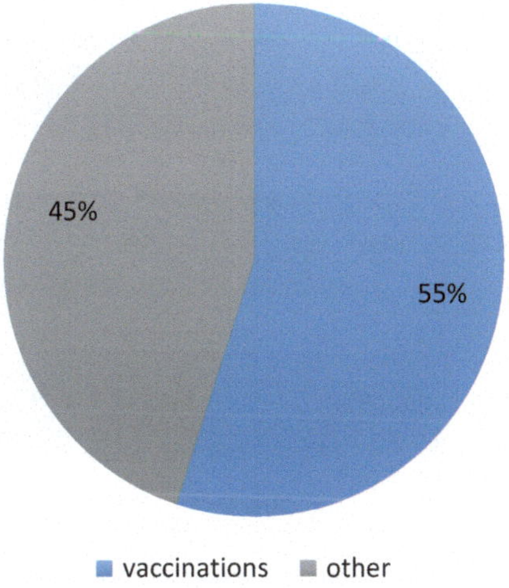

confirmed by the particular interest in issues related to Ukraine and Ukrainians and also by the emergence and rapid 'career' of issues related to the Israeli-Palestinian conflict. In terms of politics, too, fake news is reactive—it devotes attention mainly to those politicians who are also the focus of mainstream media attention. This is well known to fact-checkers, but here it has been captured not only in specific examples (case studies) but as a systemic phenomenon (Chart 5.35).

A comparison of the general themes dominating fake news in Poland, the Czech Republic, and Slovakia shows an increased interest in social issues in Poland, and a relatively small interest in health issues (Charts 5.36, 5.37, 5.38, 5.39, and 5.40). A characteristic feature of Polish fake news is also the relatively strong category varia, which refers to individual stories, life wisdom, positive heroes, etc. In the case of Slovakia, politics dominates, at the expense of social issues. The detailed analyses of the subtopics presented above also to indicate that a particularly popular topic of fake news in the Czech Republic is issues related to Ukraine—be it in the context of Zelensky, social problems or, finally, security issues.

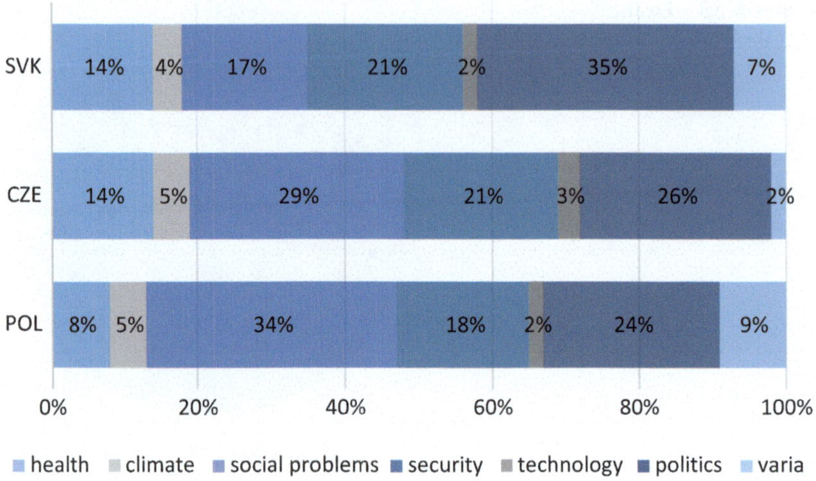

Chart 5.36 General themes—comparison

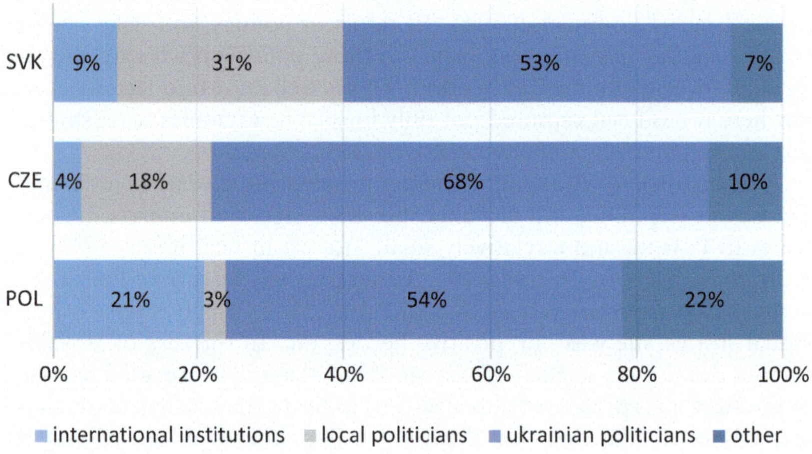

Chart 5.37 Politics—comparison

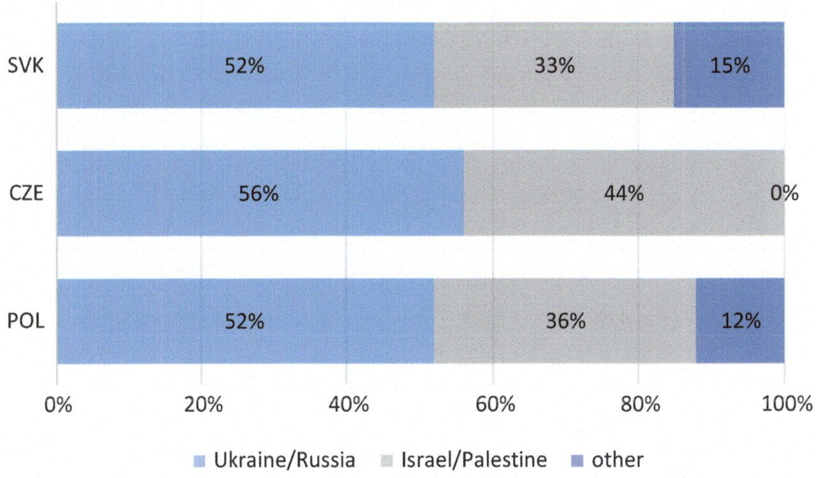

Chart 5.38 Security—comparison

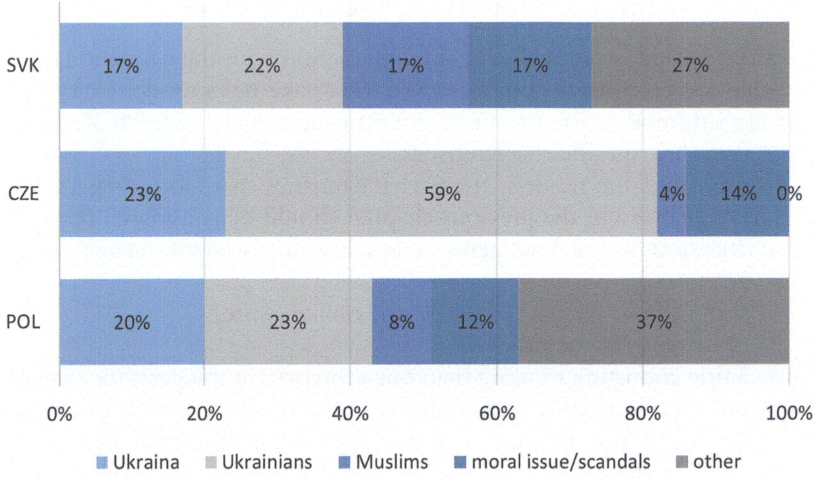

Chart 5.39 Social topics—comparison

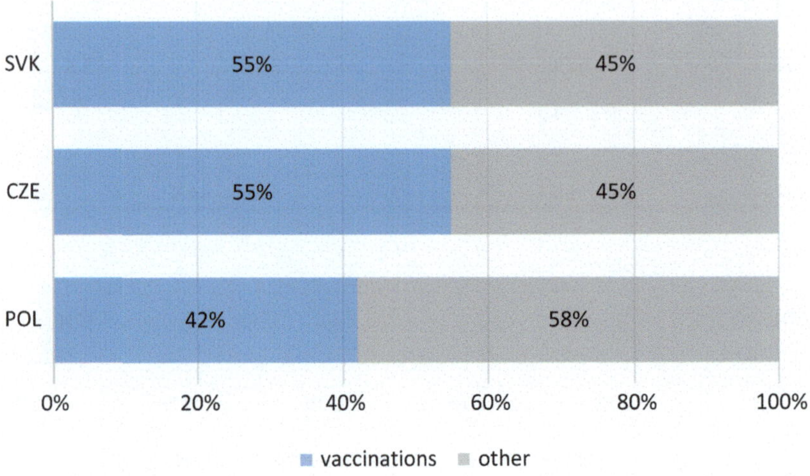

Chart 5.40 Health—comparison

5.5 Summary and Conclusions

By analysing the structural data (in conjunction with their function), we are able to reconstruct both a transnational fake news model (M1) and specific sub-models for the three countries analysed (MP; MCZ; MS), indicating their specific characteristics

In preparing the models, those characteristics that, according to the literature analysed in the previous chapter, should be considered relevant (characteristic) for the news genre were taken into account, namely:

1. text + photo format (especially the online genre);
2. occurrence of the title;
3. a title consisting of more than one sentence (in this case, the issue is not clear-cut; such a structure is undoubtedly a feature of online news, but not necessarily a feature of news in traditional media);
4. a title in the form of a single sentence or a sentence equivalent;
5. no emoticons in the title;
6. no punctuation mark (possibly a full stop) at the end of a sentence in the title;
7. no emoticons in the body of the text;
8. no capital letters in the body of the text;
9. the occurrence of figures in the body of the text;

The decision to consider the aforementioned categories as particularly relevant for news is also supported by observations on what manipulation strategies, affecting both form and semantics, occur in fake news. The most frequently mentioned are: misspelling of words; the use of capital letters and emoticons in the headline; and the use of emotionally charged words (Juez & Mackenzie, 2019).

The structure of the transnational model (M1) is presented by Chart 5.41.

The overall model of fake news overlaps with the selected features of news by 54%. It should be borne in mind, of course, that we do not have analogous figures for news, and that the features that characterise them have been determined on the basis of the literature on the subject and the possibilities of the designed research methodology. For the sake of completeness, it would be necessary to carry out analogous research on a corpus of hard news, preferably from both print and online media. Either way, it is clear that the degree to which fake news resembles news is not evenly distributed. The similarity mainly concerns sentences in the title, the lack of clear signals of the emotionality of the text such as emoticons or capital letters (especially in the title). Titles consisting of more than one sentence and figures in the text are the least frequent in the examples analysed. Note, however, that the first feature is not fully confirmed as a

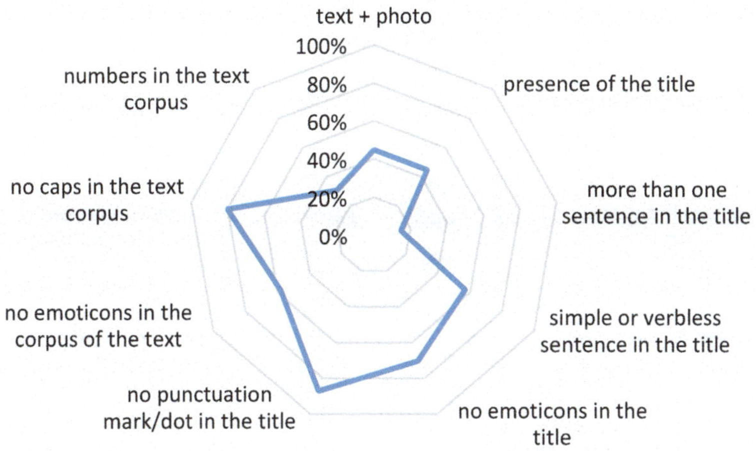

Chart 5.41 General fake news model

feature of real news, while the limited occurrence of digits may also be related to the limited function of the text in a fake news message (it is, for example, very short and the data shows a photo or a video).

Charts 5.42, 5.43, and 5.44 show the models for surveys conducted in national languages.

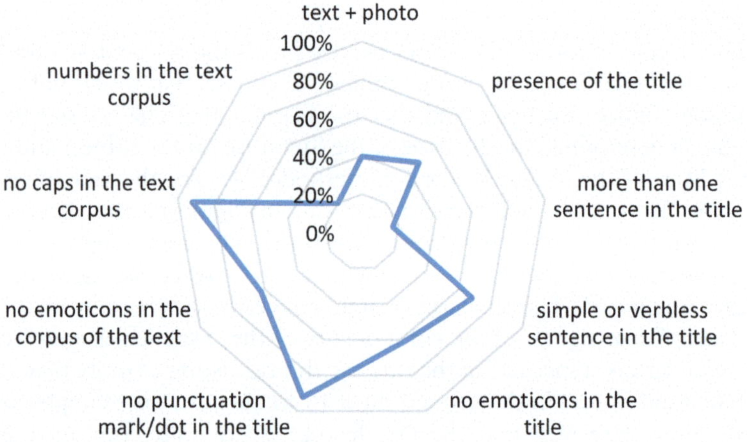

Chart 5.42 Fake news model Poland

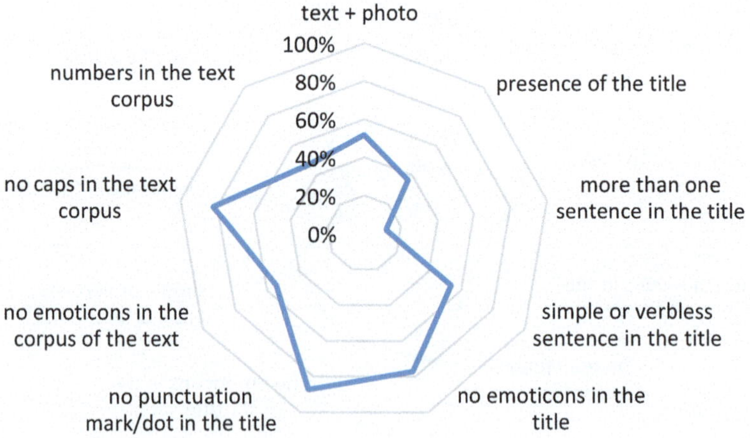

Chart 5.43 Fake news model Czech Republic

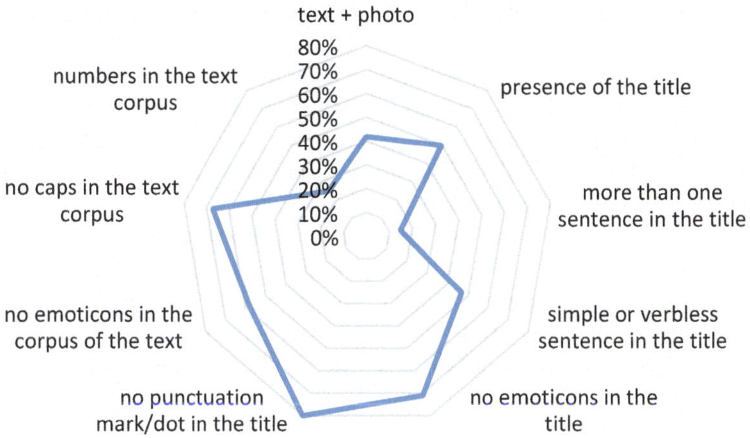

Chart 5.44 Fake news model Slovakia

In Poland, fake news is 56% similar to news. A characteristic feature is the high number of singular sentences and sentence equivalents in the title and the high predominance of titles with descriptive modality. There are very few emoticons and capital letters, but—on the other hand—also few numbers.

In Czech, fake news also overlaps with the reconstructed news model by 56%. Numbers are relatively abundant in this model (perhaps due to the fact that we observe relatively most textual messages here). There are also few emoticons and capital letters (although a relatively large number in the text corpus).

In Slovak, the fake news model overlaps with the news model to the smallest extent, although it is still over 50% (51%). In this case, there is the highest occurrence of isolated titles, which, however, are least often composed of single sentences or sentence equivalents. Capital letters are also relatively frequent in the texts.

In terms of macro-structures (themes, narratives, discourses), the similarity between fake news and news is clear. This similarity is primarily based on the similar topics that are covered. As it seems, fake news, like news, is guided by a catalogue of news values and broader criteria for the selection of presented topics, resulting from discursive practices and journalistic ideology. Of course, there is also another mechanism at work here—the direct following of what is news, what attracts public attention. In this sense, fake news is reactive, parasitic, as it were, on what is the subject matter of real news, and on the content of professional media.

Fake news, like the news, legitimises itself in its function of communicating knowledge about the world, in its factuality and objectivity primarily through the presentation of evidence, thus appealing to the field of rationality—the vision of the world as knowable and constructing the audience as capable of rational evaluation of facts.

The dominant narrative of depreciating authority figures can also be defined in this context—after all, independent and rational thinking is also critical thinking. In terms of narrative, fake news, like news, is also founded on conflict, on the disruption of the world order. Finally, like journalistic texts (news, but even journalistic investigations), it presents itself as revealing hidden knowledge about the world, especially regarding the abuse and deception of power.

The conclusions presented above are preliminary, each of the themes raised in principle calls for further analysis. First of all, at the most general level (according to the basic assumptions of discursively oriented genre research), in reconstructing the generic model of fake news, it is necessary to take into account the ways of decoding, reading texts of this nature in the context of their genre characteristics (Edgerly & Vraga, 2019). The question of the ways in which fake news is decoded, especially in the context of its impact on audiences, is the subject of much research, but research is rarely conducted in the context of the genre determinants of texts (Greifeneder et al., 2021).

Secondly, a catalogue of features considered constitutive of the news genre requires in-depth analysis. The catalogue presented in this chapter was constructed on the basis of data of a very diverse nature—from empirical data, through data from texts regulating the functioning of the professional discursive community of journalists (e.g. codes of ethics or elements of journalistic training), to abstract concepts from the field of social sciences. In the context of further research, this catalogue would need to be validated, both in terms of the qualities that are included in it and the actual occurrence of these qualities in real news. A separate analysis of the structure of online news should also be an issue.

Thirdly and finally, at a basic level, one would have to ask from what point in time can fake news be considered to mimic news in terms of genre. Are certain features more important and others less important, is it even possible to point to any numerical indicators here? Some light could certainly be shed on these issues by properly designed audience research.

In this context, it is also worth asking whether the clear thematic affinity between news and fake news is an expression of a deeper structural correlation or simply the result of fake news following news. These phenomena are not necessarily mutually exclusive, since, if we take the perspective of the viewer, similar subject matter, regardless of what motivates it, becomes a fact that sets the processes of text interpretation in a certain way.

Despite these problems, the analysis carried out points to several important issues. First of all, despite some differences, there is a relatively consistent structural model (in terms of superstructures) of fake news in the three languages (and countries) studied, which can be described in terms of selected formal features. Secondly, one can certainly speak of a tendency towards objectification in fake news. This tendency manifests itself primarily in a clear reduction of elements strongly encoding emotions, especially in titles. It is also noteworthy that almost 50% of the texts analysed have a structurally and functionally distinguishable title, which, to an even greater extent than the body of the text, is constructed in a way that suggests a move away from evaluation and emotion. In terms of macro-structures, too, similarities can be seen between the three countries (languages), accompanied, however, by quite distinctive shifts. The answer to the question of the reasons for these differences is complex—they probably have to do with the cultural and political peculiarities of the societies and countries in question, with differences in the functioning of and expectations from the media, and finally with differences in the current social and political situations. Undoubtedly, however, in terms of themes, narratives, and updated discourses (interpretations of the world) fake news in all three countries is very similar to news.

References

Alba-Juez, L., & Mackenzie, J. L. (2019). Emotion processes in discourse. *Emotion in Discourse, 302*, 1.

Allan, S. (Ed.). (2009). *The Routledge companion to news and journalism*. Routledge.

Bateman, J. A. (2008). *Multimodality and genre. A foundation for the systematic analysis of multimodal documents*. Palgrave Macmillan.

Bednarek, M., & Caple, H. (2012). *News discourse* (Vol. 46). A & C Black.

Bell, A. (1991). *The language of news media*. Blackwell.

Bell, A. (1998). The discourse structure of news stories. In A. Bell & P. Garnett (Eds.), *Approaches to media discourse*. Blackwell Publishing.

Claridge, C. (2010). News discourse. In A. H. Jucker & I. Taavitsainen (Eds.), *Historical pragmatics*. De Gruyter Mouton.
Connor, U., & Mauranen, A. (1999). Linguistic analysis of grant proposals: European Union research grants. *English for Specific Purposes, 18*(1), 47–62.
Cramer, P. A. (2011). *Controversy as news discourse* (Vol. 19). Springer Science & Business Media.
Dai, Z., Taneja, H., & Huang, R. (2018). Fine-grained structure-based news genre categorization. In *Proceedings of the Workshop Events and Stories in the News 2018*.
Davies, M. (2014). *Oppositions and ideology in news discourse*. Bloomsbury Publishing.
Deuze, M. (2005). What is journalism? Professional identity and ideology of journalists reconsidered. *Journalism, 6*(4), 442–464.
Edgerly, S., & Vraga, E. K. (2019). News, entertainment, or both? Exploring audience perceptions of media genre in a hybrid media environment. *Journalism, 20*(6), 807–826.
Fairclough, N. (1995). *Media discourse*. Arnold.
Fruttaldo, A. (2017). *News discourse and digital currents: A corpus-based genre analysis of news tickers*. Cambridge Scholars Publishing.
Gans, H. (1979). *Deciding what's news*. Vintage Books.
Greifeneder, R., Jaffe, M., Newman, E., & Schwarz, N. (2021). *The psychology of fake news: Accepting, sharing, and correcting misinformation*. Routledge.
Hall, S. (1982). The rediscovery of ideology: Return of the repressed in media studies. In M. Gurevitch, T. Bennet, J. Curran, & J. Woollacott (Eds.), *Culture, society and the media*. Methuen.
Halliday, M. A. K. (2003). *On language and linguistics* (Vol. III). Continuum.
Hartley, J. (2005). To attract or to inform: What are titles for? *Journal of Technical Writing and Communication, 35*(2), 203–213.
Heppell, F., Bontcheva, K., & Scarton, C. (2023). Analysing State-Backed Propaganda Websites: A New Dataset and Linguistic Study. In Proceedings of the 2023 Conference on Empirical Methods in *Natural Language Processing* (pp. 5729–5741). Singapore. Association for Computational Linguistics.
Horne, B., & Adali, S. (2017). This just in: Fake news packs a lot in title, uses simpler, repetitive content in text body, more similar to satire than real news. In *Proceedings of the international AAAI conference on web and social media*, Vol. 11 (1), 759–766.
Humprecht, E. (2019). Where 'fake news' flourishes: A comparison across four Western democracies. *Information, Communication & Society, 22*(13), 1973–1988.
Igwebuike, E. E., & Chimuanya, L. (2020). Legitimating falsehood in social media: A discourse analysis of political fake news. *Discourse & Communication, 15*(1), 42–58.

Ihlström, C. (2004). *The evolution of a new(s) genre.* Gothenburg Studies in Informatics, Report 29.

Isani, S. (2011). *Of headlines & headlinese: Towards distinctive linguistic and pragmatic genericity.* ASp [Online], 60, 81–102.

Juez, L. A., & Mackenzie, J. L. (2019). Emotion, lies, and 'bullshit' in journalistic discourse. *Ibérica, 38*, 17–50.

Jufri, J., & Usman, U. (2023). Issues of the 2024 presidential candidates in online news texts: Critical discourse analysis of the Teun A Van Dijk model. *Journal of Asian Multicultural Research for Social Sciences Study, 4*(1), 1–26.

Kapusta, J., Hájek, P., Munk, M., & Benko, Ľ. (2020). Comparison of fake and real news based on morphological analysis. *Procedia Computer Science, 171*, 2285–2293.

Lazer, D. M., Baum, M. A., Benkler, Y., Berinsky, A. J., Greenhill, K. M., Menczer, F., et al. (2018). The science of fake news. *Science, 359*(6380), 1094–1096.

Maci, S. M. (2019). Discourse strategies of fake news in the anti-vax campaign. *Lingue Culture Mediazioni-Languages Cultures Mediation (LCM Journal), 6*(1), 15–43.

Molek-Kozakowska, K. (2013). Towards a pragma-linguistic framework for the study of sensationalism in news headlines. *Discourse & Communication*, 7(2), 173–197. https://doi.org/10.1177/1750481312471668

Molek-kozakowska, K. (2014). Coercive metaphors in news headlines: A cognitive-pragmatic approach. *Brno studies in English, 40*(1), [149]–173.

Mu, Y., Jiang, Y., Heppell, F., Singh, I., Scarton, C., Bontcheva, K., & Song, X. (2023). A large-scale comparative study of accurate COVID-19 information versus misinformation. arXiv preprint arXiv:2304.04811.

Raza, S., & Ding, C. (2022). Fake news detection based on news content and social contexts: A transformer-based approach. *International Journal of Data Science and Analytics, 13*(4), 335–362.

Reese, S. D. (1990). The news paradigm and the ideology of objectivity: A socialist at the Wall Street Journal. *Critical Studies in Media Communication, 7*(4), 390–409.

Rubin, V. L., Conroy, N. J., & Chen, Y. (2015). Towards news verification: Deception detection methods for news discourse. Hawaii International Conference on System Sciences (HICSS48) Symposium on Rapid Screening Technologies, Deception Detection and Credibility Assessment Symposium, January 5-8, 2015 Grand Hyatt, Kauai]In.

Sala, M. (2023). The expression of bias in (online) newspaper headlines. In *The Routledge handbook of discourse and disinformation.* Routledge.

Schudson, M. (1978). *Discovering the news: A social history of American newspapers.* Basic Books.

Soloski, J. (1989). News reporting and professionalism: Some constraints on the reporting of the news. *Media, Culture & Society, 11*(2), 207–228.

Song, Y., & Lin, A. (2012). A genre-aware approach to online journalism education. *Procedia – Social and Behavioral Sciences, 51*, 400–404.
Stasiuk-Krajewska, K. (2018). *Dziennikarstwo i media. Struktury dyskursu i hegemonia*. Elipsa.
Stasiuk-Krajewska, K. (2024). Fake news as text and genre. Selected linguistic (communicative) structures and their functions. In K. Stasiuk-Krajewska, M. Wenzel, & K. Bąkowicz (Eds.), *Disinformation. Social and media context*. Peter Lang.
Swales, J. M. (1990). *Genre analysis: English in Academic and Research Settings*. Cambridge University Press.
van Dijk, T. A. (1983). Discourse analysis: Its development and application to the structure of news. *Journal of Communication, 33*(2), 20–43.
van Dijk, T. A. (1985). Structures of news in the press. In T. A. van Dijk (Ed.), *Discourse and communication: New approaches to the analysis of mass media discourse and communication*. Walter de Gruyter.
van Dijk, T. (1988). *News as discourse*. Lawrence Erlbaum Associates.
van Dijk, T. A. (2019). *Macrostructures: An interdisciplinary study of global structures in discourse, interaction, and cognition*. Routledge.
van Leeuwen, T. (2007). Legitimation in discourse and communication. *Discourse & Communication, 1*(1), 91–112.
van Leeuwen, T. (2008a). *Discourse and practice: New tools for critical discourse analysis*. Oxford University Press.
van Leeuwen, T. (2008b). News genres. In R. Wodak & V. Koller (Eds.), *Handbook of communication in the public sphere*. Mouton de Gruyter.
White, P. (1997). Death, disruption and the moral order: The narrative impulse in mass-media hard news reporting. In F. Christie & J. R. Martin (Eds.), *Genres and institutions: Social processes in the workplace and school*. Cassell.
Yarlott, W. V., Cornelio, C., Gao, T., & Finlayson, M. (2018). Identifying the discourse function of news article paragraphs. In *Proceedings of the Workshop Events and Stories in the News 2018. Association for Computational Linguistics*.
Zelizer, B. (2004). When facts, truth, and reality are God-terms: On journalism's uneasy place in cultural studies. *Communication and Critical/Cultural Studies, 1*(1), 100–119.

CHAPTER 6

Visual Structures of Disinformation (Fake News) in Poland, Czech Republic, and Slovakia

Karina Stasiuk-Krajewska

6.1 Introduction

This chapter addresses issues of visuality (static images) in fake news. The findings from Chaps. 2 and 5, concerning the notion of genre and news as a genre, remain in force. This time, however, the issue under examination and comparative analysis was that of the images that accompany most of the disinformation messages analysed. The context of the research and the analysis of its results remain the news genre, this time intepreted through the prism of news values. As I will show in the following section, the category of news values correlates very well with the discursive analysis of news as a genre. The category will therefore be examined in the context of so-called constructivist definitions of journalism, multimodal discourse analysis and selected aspects of journalistic photography.

K. Stasiuk-Krajewska (✉)
Faculty od Law and Communication, SWPS University, Wrocław, Poland
e-mail: kstasiuk-krajewska@swps.edu.pl

M. Wenzel, K. Stasiuk-Krajewska (eds.), *Disinformation in Central Europe*, https://doi.org/10.1007/978-3-031-76871-2_6

6.2 Theoretical Context

In the 1970s, Gaye Tuchman (Tuchman, 1972, 1978), based on her ethnographic research on news production, concluded that contemporary news theory (in the United States) developed with the American social structure and is therefore inextricably linked to capitalism and democracy. News, being the most typical and highly valued journalistic product, is therefore not a reflection of reality, but is historical and social in nature, a method of constructing the social world as it shapes and defines, rather than reports, the events described. Put differently, news constitutes a method of framing the experience of everyday life. As Michal Schudson points out "Framing is as central a concept as there is in the study of news. It moves the analysis of news away from the idea of intentional bias. That is to acknowledge that news stories frame reality and also to acknowledge that it would be humanly impossible to avoid framing. Every narrative account of reality necessarily presents some things and not others; consciously or unconsciously, every narrative makes assumptions about how the world works, what is important, what makes sense, and what should be" (Schudson, 2003: 35–36). Indeed: "News is a window on the world. Through. [...] But, like any frame that delineates a world, the news frame may be considered problematic. The view through a window depends upon whether the window is large or small, has many panes or few, whether the glass is opaque or clear, whether the window faces a street or backyard" (Tuchman, 1978: 1).

This point can be made even more broadly by discussing the connection of news not only to American society, but also to the broader cult-historical formation (discursive field, Laclau & Mouffe, 1985), that is modernism, as mentioned earlier in Chap. 5. As John Hartley writes, "news is 'the sense-making practice of modernity' and at the same time 'the most important textual system' in the modern world" (Hartley, 1996: 32). This kind of thinking about news has also been referred to by W. Lance Bennett, who emphasised that news is an element of political illusion, a social construct created in several stages: promotion by politicians, production in media organisations (as Schudson puts it: "News, [...], like bread or a sausage, is something people make", 2003: 4), technology, and finally—selling news as a consumer commodity (Bennett, 2016). News is thus, on the one hand, a product of the historical situation in which it is produced, but on the other hand, it itself also contributes to the specificity of this reality, the specificity of defining the social world.

Indeed, what has been said about news can be said about all social practices that are discursive in nature, that is, mediated by signs and involved in making sense of the experience of the world. In this sense, the position presented here does not have to be a criticism of journalism or an appeal for journalists to go beyond this cognitive or discursive framework. In fact, it cannot be such an appeal, because in the constructivist view, leaving the discursive framework is simply not possible. This is due to the fundamental characteristic and function of language, which is not a neutral tool for describing the world, but is always its construction.

However, the issue is not so simple. After all, journalists, as a professional group, have a special social position, they legitimise their activities by appealing to specific values. And news, of necessity, is linked either to the practice of journalism understood as a discursive practice (Tuchman, 1972, 1978; Patterson, 2013) or to the discourse of journalism, often referred to as journalistic ideology. An ideological construction, presenting a culture's prevailing beliefs about reality (Allan, 2010), but also constructing a framework of public experience, a belief about what is real and what is relevant (Schudson, 2003). News, as a genre for which journalists have discursive authority, thus becomes the basis of the "newsmen's occupational ideology" (Schlesinger, 1978), "ideology of professionalisation" (Soloski, 1989), "journalists' professional ideology" (Zelizer, 2004) understood as a kind of "collective knowledge" of the journalistic community. The ideology is generally defined as the dominant way in which journalists evaluate the news and give meaning to their work, the following assumptions are usually considered to be its essential elements: journalists provide a public service; journalists are neutral, objective, fair and (thus) credible; journalists must enjoy editorial autonomy, freedom and independence; journalists have a sense of immediacy, finally journalists have a sense of ethics and legitimacy (Deuze, 2005).

Understood in this way, journalistic ideology has several important functions for this group: it is a "strategic ritual" for positioning oneself in the profession vis-à-vis media critics and the public (Tuchman, 1972); a tool in the hands of journalists and editors to legitimise the structure of the news organisation or media corporation for which they work (Soloski, 1989); and finally, a means of justifying the choices made and defending themselves against criticism. Interestingly, and which is also confirmed by research on news values (as discussed below), the move of journalistic news activity to the Internet has not led to fundamental changes in journalistic professional ideology, with the possible exception of a greater emphasis on transparency due to increased opportunities for audience control (Deuze, 2005, Hall et al., 1978).

The above-referenced constructivist approach to news and journalism sheds a specific light on news values. In the most general theoretical frame, news values are an essential element (next to e.g. individual bias, stereotypes or external social or economic conditions) of the production and maintenance of journalistic ideology. They are "an ideological code" (Hartley, 2013), presenting the world in a very specific way. For Stuart Hall (1973), although the news values of mainstream journalism are ostensibly a set of neutral, routine practices, they are in fact part of an "ideological structure", a socially determined constrution of reality that privileges the perspectives of the most powerful social groups. This fact, as the author notes, is often not even realised by journalists themselves, who proclaim with absolute conviction and without a shadow of cynicism their belief in their objectivity and work for the social good. Edward Herman and Noam Chomsky (2010) go further, suggesting that 'topic selection' is one of the key ways in which the media fulfils its 'social purpose' of instilling the economic, social, and political agenda of the groups that dominate society and the state. Similarly, Fowler (2013), Bell (1991) and Cotter (2010) have written about news values as an ideology or referred to them as an ideological factoid.

What is the realisation of this ideology mediated by news values? The news media select events to report according to a complex set of criteria; therefore, news is not simply what happens, but what can be considered and presented as newsworthy because of news values. As stated by Hall, "Things are newsworthy because they represent the changefulness, the unpredictability and the conflicting nature of the world. But such events cannot be allowed to remain in the limbo of the 'random'—they must be brought within the horizon of the 'meaningful'. This bringing of events within the realm of meaning means, in essence, referring unusual and unexpected events to 'maps of meaing' which already form the basis of our cultural knowledge, into which the social world is already 'mapped'" (Hall et al., 1978: 54). From a cognitive point of view, news values can be regarded as intersubjective mental categories. In determining the meaning of events, media personnel and audiences refer, most often implicitly and unconsciously, to what in cognitive psychology and semiotics are called 'frames', 'paradigms', 'stereotypes', 'schemas' or 'generalised assumptions'.

The most typical definition of news values essentially considers them as criteria that guide journalists in their work (hence their obvious connection to journalistic ideology). In this view, news values are defined as "properties of events or stories or as criteria/principles that are applied by news workers in order to select events or stories as news or to choose the structure and order of reporting" (Bednarek & Caple, 2014: 2). The link to journalistic ideology also stems from the fact that news values are located as well in the field of social cognition and as such are a discursive phenomenon. van Dijk (1988b) underlines their location in social cognition, as they are values about the newsworthiness of events or discourse, shared by professionals, but also by the public of the news media.

Although the idea of news values as a useful category to describe journalism and the media dates back to Walter Lippmann (2020), the first set of news values was described by Galtung and Ruge in 1965 and included 12 factors, which can be briefly presented as follows (description of values after: Wahl-Jorgensen & Hanitzsch, 2020):

1. Frequency: an event that unfolds within a publication cycle of the news medium is more likely to be selected than one that takes place over a long period of time.
2. Threshold: Events have to pass a threshold before being recorded at all; the greater the intensity (the more gruesome the murder or the more casualties in an accident), the greater the impact and the more likely it is to be selected.
3. Unambiguity: The more clearly an event can be understood and interpreted without multiple meanings, the more likely it is to be selected.
4. Meaningfulness: The culturally familiar is more likely to be selected.
5. Consonance: The news selector may be able to predict (due to experience) events that will be newsworthy, thus forming a 'pre-image' of an event, which in turn increases its chances of becoming news.
6. Unexpectedness: Among events meaningful and/or consonant, the unexpected or rare event is more likely to be selected.
7. Continuity: An event already in the news has a good chance of remaining in the news (even if its impact has been reduced) because it has become familiar and easier to interpret.
8. Composition: An event may be included as news less because of its intrinsic news value than because it fits into the overall composition or balance of a newspaper or news broadcast.
9. Reference to elite nations: The actions of elite nations are seen as more consequential than the actions of other nations.

10. Reference to elite people: Again, the actions of elite people, likely to be famous, may be seen by news selectors as having more consequence than others, and news audiences may identify with them.
11. Reference to persons: News that can be presented in terms of individual people rather than abstractions is likely to be selected.
12. Reference to something negative: Bad events are generally unambiguous and newsworthy.

Currently, there are very many approaches and catalogues of news values in the literature, which are generally quite similar to each other (short literature review on news values: Harcup & O'Neill, 2017). The presentation of these catalogues is not necessary in the context of this text, let us just cite Bell's catalogue of news values (1991), which, based on the original assumptions of Galtung and Ruge, indicated that news values can operate on three levels:

1. values in the news text;
2. values in the news process;
3. values in news actors and events.

This approach is interesting because it puts the category of news values in a very broad context, it also makes it possible to propose a solution to a certain inconsistency between the expectations of news, which stem from the perception of journalism as objective, oriented towards socially responsible communication of knowledge about the world, and the discourse of news values, where expectations generally refer to how to construct texts that will be attractive to the audience (Stasiuk, navigator).

Values in the news text are considered by Bell as factors that build the quality of the news text, including clarity, brevity, and colour. Bednarek and Caple (2012a) refer to them as "news writing objectives" (Bednarek & Caple, 2012a). In turn, Bell lists the following as values in news processes: continuity; competition; co-option; composition; predictability; and prefabrication. As can be seen, these factors are mostly related to aspects external to, but clearly influencing, the structure of journalistic texts. For Caple and Bednarek, these are "news cycle/market factors" (Bednarek & Caple, 2012a). Only the third category, values in news actors and events, leads to what could be termed 'cognitive' news values. These include: negativity; recency (or timeliness); proximity; prominence/eliteness; consonance; impact; novelty; superlativeness; facticity.

Only two examples of news values catalogues have been presented above, although, as already mentioned, the category of news values has received a great deal of theoretical elaboration and operationalisation in research and analysis. Marek Palczewski (2018: 79–84) analysing the history of the presence of this concept in media studies, lists 79 news values, originating from 40 classifications. Hence, some authors aim to limit the catalogue of news values, pointing to such values that should be considered basic (for example: the presence of conflict or deviance, relevance to the audience, a sense of cultural affinity, and degree of prominence within the hierarchy of nations (Golan, 2008). Hartmann and Husband (1976), on the other hand, point to two characteristics that make events newsworthy: conflict and the potential to be interpreted within a safe and recognisable range of worldviews and stereotypes. Note that similar characteristics are pointed to as reasons for the popularity and strong impact of fake news. There are also many attempts in the literature to structure news values, for example Winfried Schulz (1982), carrying out a content analysis of newspapers, proposed six different dimensions to news selection, which he further broke down into 19 news factors: status (elite nation, elite institution, elite person); valence (aggression, controversy, values, success); relevance (consequence, concern); identification (proximity, ethnocentrism, personalisation, emotions); consonance (theme, stereotype, predictability); and dynamics (timeliness, uncertainty, unexpectedness).

Whether and how news values have been modified by the proliferation of online journalism is also the subject of analysis. In this context, a slightly modified catalogue of news values has been proposed, which includes: exclusivity; bad news; conflict; surprise; audio-visuals; shareability; entertainment; drama; follow-up; the power elite; relevance: stories about groups or nations perceived to be influential with, or culturally or historically familiar to, the audience; magnitude; celebrity; good news; the news organisation's agenda (Harcup & O'Neill, 2017). The proposal seems to need some sorting out, as it concerns both the cognitive, organisational and formal (textual) determinants of news values. The relationship between "traditional news values" and those identified as new in the context of online communication should also be carefully analysed (e.g. to what extent is shareability a result of other news values and to what extent is it a separate value).

An attempt to structure the theory of news values leads first of all to the identification of three basic approaches to the phenomenon (Bednarek, 2006). Firstly, news values are seen as external to the text, for example as values that operate in the minds of journalists, or related routines and procedures, or, even more differently, as a system of criteria that guide journalists' decision-making. Stewart Hall's theory, discussed earlier, for example, is located in this view.

The second perspective, termed material, locates news values in events, makes them features external to journalism, coming from objectively existing reality. These features cause the media to take an interest in a given event. This approach does not occur in a "pure" state (for then it would not be connected to media theory or journalism in any way), and 'the identification of news values in stories offers insights into the how of the reporting rather than the why' (Bednarek & Caple, 2017). However, many authors mention that releases have a certain "instristic newsworthiness" (Harcup & O'Neill, 2001).

The third perspective, the most interesting in the context of the research presented here, relates to how news values are constructed in the text, as its semiotic resources. I will return to its specific assumptions later in the chapter.

The questionable theoretical status of the second view of news values leads to a simplified classification, which includes a 'cognitive' view of news values and a discursive view of them. From a cognitive perspective, news values are conceptualised as beliefs (or criteria); from a discursive perspective, in terms of how news is constructed through discourse (both language and visuality) (Bednarek & Caple, 2012b). Let us note in passing that, given the assumption of a discursively oriented theory of journalism and news as discussed in the second chapter of the book, these approaches do not contradict each other—for the construction of a text is an effect of the rules of discourse. A closer analysis of news values as a discursive phenomenon indicates that one should rather speak of the dimensions of news values. For example, van Dijk points out that they have cognitive (are culturally "shared mental objects of social cognition"), social (are "applied by social members in a large variety of practices and contexts") and discursive aspects (van Dijk, 1988a). A similar view is presented in the Table 6.1 below.

From the point of view of the genre analysis of fake news in the news context, it is particularly relevant that news values are sometimes considered to be constitutive of news as a genre, for example:

Table 6.1 Dimensions of news values

Material: an event in its material reality holds potential news value in a given community	*What are an event's potential news values?*
Cognitive: news workers and audience members have beliefs about news values and newsworthiness	*What beliefs do news workers and/or audience members hold about news values?*
Social: news values are applied as selection criteria in journalistic routines and practices	*How do news workers apply news values as criteria in selecting what events to cover, publish and in how to produce them?*
Discursive: news values can be communicated through discourse	*How are news values communicated through discourse, pre-, during, and post-news production and in news products?*

Source: Bednarek and Caple (2017: 43)

- "news: important information delivered in timely fashion to people who want to know it" (Bennett, 2016: 3);
- news is "judged to be newsworthy by journalists, who exercise their news sense within the constraints of the news organisations within which they operate" (Harrison, 2006: 13).

As such, they also have a discursive or ideological dimension and provide a framework for constructing news.

Much of the above-mentioned catalogues of news values have emerged from studies of media messages, most often, however, related to the analysis of a limited number of cases, qualitative rather than quantitative in nature. More recently, concepts and studies based on them have also emerged attempting to characterise news values in journalistic texts by referring to linguistic analysis and using large text corpora and CADS: Corpus Assisted Discourse Studies (Bednarek & Caple, 2014). Such research is based on the assumption, extensively discussed in earlier chapters, that the semiotic tools that construct news values in texts are highly conventionalized in journalistic practice, and that the preparation of a semantic inventory of such elements will make it possible to specify the rather general, as already discussed, scopes of individual news values and especially their functioning in journalistic practice (Bednarek, 2006). Such an approach seems all the more justified given that many publications emphasise the mediation of news values by language (which is not surprising if we are talking about a genre that uses linguistic structures and semantics). In this context, the relationship between news values and language is sometimes referred to as: linguistic representations/features, contributing to, increasing, conforming to, stressing or being linguistic manifestations of particular news values (cf. Bednarek & Caple, 2014). Linguistic elements that could be analysed in the context of news values studies are mainly: evaluative language, Intensification and quantification, comparison, references to emotion, "negative" vocabulary, word combinations, metaphor/simile, story structure, references to time and place, references to the nation/community, first-person plural pronouns, role labels, quotes from "ordinary" individuals, references to individuals (Bednarek & Caple, 2012b: 46–47). A full discussion of linguistic resources for establishing news values, together with an example of the use of quantitative methods in an extended research project, can be found in: Bednarek and Caple (2017).

This kind of research seems very interesting, but poses the difficulties mentioned in the previous chapters: related to the construction of a large corpus of data and the transition from lexical analysis to semantic analysis, which would allow a clear separation of the ways in which news values are encoded. From the point of view of the smaller chapter, the fundamental problem is, of course, that this kind of approach cannot be applied to the study of visual material, although the desire to organise and catalogue linguistic features may be an important inspiration for addressing a similar challenge in the context of visually encoded semantics.

The issue of visuality in the media (including news and fake news) is important primarily because of the increasing role of the image in news messages. There are studies on the relationship between visual elements and linguistic elements in news, but these are primarily concerned with television broadcasts (Meinhof, 1994). Pioneering in this respect is Caple and Bednarek's research, which analyses both linguistic and visual tools for encoding news values, but, as the authors acknowledge: "New ways of telling stories online, such as on news homepages or in multimedia sections, also incorporate both verbal and visual elements. The organisation of such verbal-visual texts and the relations between the verbal and the visual thus deserve further attention" (Caple & Bednarek, 2013: 112).

An adequate theoretical (and to some extent methodological) background for this kind of research and analysis is multimodal discourse analysis (MDA) and, referring to it, multimodal genre analysis.

MDA is associated with the so-called visual turn in discourse studies (within critical discourse analysis). It is mainly inspired by the work of

Gunther Kress and Theo van Leeuwen. Kress and van Leeuwen showed that much communication is "multimodal" rather than "monomodal". At a general level, multimodal discourse analysis seeks to provide an integral and coherent picture of multimodal communication and all its resources, also in the context of ways to integrate tchem.

On the basis of the assumptions of multimodal discourse analysis grew social semiotics, whose tools are quite often used in the analysis of visual messages (Hodge & Kress, 1998; van Leeuwen, 2005; Kress & van Leeuwen, 1996/2006, 2002; Chouliaraki, 2012; Caple & Bednarek, 2017). Social semiotics seeks answers to the question of how meaning is realised in visual processes, using concepts such as composition, modality or framing. It thus analyses how discourses, myths or ideologies are constructed and transformed in visual communication processes. Social semiotics involves a method referred to as the "visual grammar", which is built on the assumption that, like linguistic structures, visual structures point to specific interpretations of experience and forms of social interaction, because meanings belong to culture rather than to specific semiotic modes (Kress & van Leeuwen, 1996/2006). Another interesting example of the operationalisation of the assumptions of social semiotics is the 8 modality scales, which make it possible, for example, to determine the degree of reduction in an image (which translates into its abstractness), the degree of ambiguity (linked to the belief in the veracity of the image) or the degree of articulation of depth and modulation of colour (Kress & van Leeuwen, 1996/2006).

Within the framework of multimodal discourse analysis, researchers are not only interested in visuality, but also in other ways of constructing meaning, such as sound or music (van Leeuwen, 1999, 2005; Kress & van Leeuwen, 2010, 2001). The assumption here is that societies have diverse forms of representation with specific meaning-making potentials in specific social communicative contexts. Differentiated meaning-making potentials lead to differentiated forms of discourse production and—consequently—open up the possibility for the social actor to use multiple forms of meaning processing in the processes of constructing individual subjectivity. At the same time, these forms do not occur separately, they coexist in all the processes described above (Manghani et al., 2006). Multimodal discourse analysis thus concertedly focuses on a peculiar network of semantic choices that are encoded by diverse sign resources. The question it poses, however, is a functional one: how such complex semantic systems perform their communicative, and therefore discursive, function—how they construct a relatively coherent interpretation of the social world (Bateman, 2011). Thus, the analysis of visual messages is at the same time a study of their function, for elements with similar discursive structures are also functionally identical (Lemke, 1999).

In the context of multimodal discourse analysis, the category of multimodal genre has emerged, referring to genres that involve the use of multiple communicative modes. Multimodal genre analysis involves, in general, applying the tools of semiotic and discourse analysis discussed in the second chapter of the book to that which is multimodal, i.e. signs that are not only linguistic in nature (van Leeuwen, 2005). In the analysis of multimodal genres, essentially the same assumptions that were mentioned in the context of the analysis of news as a genre apply, thus emphasising both its social and discursive character (and, in the context of the latter, its dependence on the specificity of the codes the genre uses, the medium of transmission or technological constraints) (Bateman, 2011).

In the context of news research inspired by multimodal discourse analysis, static visuality remains the main focus, in addition to the still dominant text (which is also related to the structure of news, within which video or separately treated sound is still rare). In this context, either specific visual realisations, images appearing in the news, such as maps or photographs, or broader structures, such as front pages, are described. Also analysed are the ways in which specific events are visually presented, such as the Israeli-Palestinian conflict. Media researchers are particularly interested in the visual aspects of online journalism (Zelizer, 2004).

However, as Barbie Zelizer (2004) notes, images are still relatively marginalised in journalism research. Meanwhile, the role of visual messages (photographs) in news is very important, also considering the mechanisms of image reception. Numerous studies indicate that images evoke stronger physiological responses than textual messages—visual perception is accompanied by a state of internal arousal or alertness, which in turn leads to increased depth of mental and emotional processing. Thus, visualisations are encoded in memory more easily, quickly and deeply and remembered for longer than texts. Once visual attention is attracted, images are perceived by a "parallel processing system", which enables the rapid perception of large and multifaceted chunks of information. The bandwidth of perceived information includes both emotional and cognitive information processing, seamlessly integrating both. Moreover, due to the holistic and rapid mode of perception, visual perception is initially subject to little conscious cognitive control. Visual messages do not have a formalised logical structure that enforces a hierarchical order of perception. Meaning is produced here not linearly, but "holistically" so to speak. In addition, visual information has a specific semiotic quality that makes it possible to encode and decode information as concrete representations rather than through abstract signs. Consequently, visuals are perceived in the same way as phenomena in the real world (more: Stasiuk-Krajewska, 2024).

In addition, in most cases, in the context of analyses devoted to journalism, images have been described as truth-telling, objective, as mirrors of the events they depict, and the idea that pictures can rather evoke emotions has been treated as a certain kind of disturbance, an irritation of the system (Zelizer, 2010). Meanwhile, pictures, by their very semiotic nature, cannot perform a purely informative function because "first, pictures function in ways that have little to do with a definitive set of certain and unambiguous cues for understanding the world: they are thought to be analogic-continuous and operating in more/less terms-rather than digital-discrete and operating in either/or terms. Generally offering an affective and often gestalt-driven view of the world, they tend to be indexical-directing attention to something; material- having a tangible form; iconic and syntactically indeterminate-representative of something but in a fuzzy, porous way. Most importantly, images are expected to offer only fragments of understanding, and thus direct their viewers elsewhere to understand what is shown-to the purposes, processes, and formulations at work beyond a camera's frame. "While words are valued for their evidentiary qualities, images offer instead implicative relays, suggestive slices of action that people need to complete by interpreting and imagining what unfolds beyond the camera's frame" (Zelizer, 2010: 6).

This observation follows directly from Roland Barthes's theory of photography (Barthes, 1977) within which he speaks of denotation and connotation, which in turn leads to the actualisation of myth, a phenomenon essentially akin to discourse, explaining the world unequivocally. However, in the context of the study of photography, the relationship between the first two levels of signification—the aforementioned denotation and connotation—is particularly interesting. Denotation is, for Barthes, what we might call the basic meaning of an image, the answer to the question: what is seen? Although even denotation is not freed from elements of cultural interpretation, at the second level of signification, at the level of connotation, this interpretation acquires special meanings. The images, subjected to the photographer's intentional processing, become, namely, a carrier, a catalyst for prejudices, stereotypes, emotions, or judgements. They are not, therefore, an objective representation of reality; for the rest, as semantic systems, they cannot be. Put differently, denotative meanings are expressed through what is depicted in the image (participants, actions, circumstances). The second level, connotative, is described as the art or processing of the image that leads to orienting its interpretation in the context of the values of a given culture.

However, most journalists, at least on a declarative level, still consider as more professional those photos that appear realistic, seeming to reflect the truth, ignoring or rejecting the existence of a level of connotation (let us note in passing that this is a journalistic 'everyday occurrence', the case is different with photos awarded at, for example, the World Press Photo competition). Moreover, "adopting the sentiments of most journalists, complementary research strands on news production, content, and effects have tended to position news images in a supportive role to words, where the verbal record underpinning journalists' authority as arbiters of the real world takes precedence over its visual counterpart. Journalism, largely seen as a project of modernity, is presumed driven by words. Though images relate variously to them-bolstering, complementing, negating, and affirming what they stipulate-words remain journalism's authoritative cues because they are thought to enable information, evidence, reasoned testimony, and deliberation" (Bednarek & Caple, 2014, p. 3).

Meanwhile, the themes related to the semantics and function of visuality obviously also apply to journalistic photography or the broader phenomenon known as visual journalism (Mirzoeff, 2002). Photographs present in the media function as symbolic representations of places, social groups, institutions or events, and in this sense they are obviously discursive in nature. At the same time, it is worth emphasising that, although the image in the media can and does perform many diverse functions (Bednarek & Caple, 2012b), contrary to professional self-description, the informational, illustrative function is by no means dominant—the most famous images, iconic photos, ideological memes, going beyond the representation of specific events, but symbolising certain ideas, values or even myths).

The reality principle ("the idea that images can 'show' or otherwise provide some valid or true information, and that vision is related to authoritative knowledge") thus seems increasingly secondary to the pleasure principle ("the idea that what really matters is not the information provided by the image and the relationship between the image and what it shows, but rather the ritualistic use of the image in constituting a community of meanings and ultimately a relationship between individuals united by the fact of giving the same meaning to the same image") or the *hyper-reality* principle (the idea that the image, rather than providing information or performing a ritual, "can be a simulacra: a communicative tool used to 'hide' rather than 'show' relevant aspects of reality" (Stocchetti, 2017).

In view of the aforementioned focus of journalistic ideology on photography in the news function (possibly treating it as a means of attracting and retaining the reader's attention), it is not surprising that the question of news values in the context of the visual has not been of particular interest to either the journalistic community or media researchers. Meanwhile, given that "images do not simply record an objective truth-rather, they offer one visual representation of an event through the ways in which elements, contexts, and participants are framed, cropped (in or out), or angled within the image frame" (Bednarek & Caple, 2017: 52), one must also assume that they are certainly carriers of news values.

Essentially in this context, it is worth mentioning Stuart Hall's conception of the news photograph to include the unexpected, dramatic, recent event concerning a person of high status (Hall, 1973). Hall locates the consideration of news values in visual messages in the aforementioned context of professional news ideology. Taking inspiration from Roland Barthes' conception, Hall points out that there are two levels of visual news meaning. The first involves the formal news value of the photographic sign, thus elaborating the photograph and text in terms of professional news ideology (news values); the second level of meaning is the ideological level of connoted themes and interpretations (Hall, 1973). In this context, Hall points out that, in fact, the connotation of a photograph is much more important than its denotation; moreover, photographs accompanying news are often used in such a way that they refer not to the essence of the information conveyed, but to its peripheral elements, to what is 'mythical', what arouses emotions and actualises stereotypes. It is significant that Hall does not treat the image as subordinate to the text or dependent in its occurrence or otherwise on the subject matter of the text (Harcup & O'Neill, 2001) but as a semiotic entity in its own right.

A full study of news values in the context of news photography is brought by an approach termed Discourse News Values Analysis (DNVA) by Bednarek and Caple (mainly: Bednarek & Caple, 2012b, 2017). Discursivity in this approach means focusing on how news values are constructed in the text through diverse elements of a semiotic nature, including visual elements. The category of discourse here is, analogous to the approach to the analysis of fake news presented in chapter two of this book, understood as a set of semiotic resources (language, image, typography, layout, sound etc.) that are involved in the construction of a text (published news story). Discourse is also a specific communicative practice that, through semantics and structures, produces and sustains certain interpretations of reality (hence the reference to constructivist interpretations of news and journalism is relevant in this view). Discourse is multimodal in nature, so it is not limited to visual elements only, but also incorporates other types of codes, especially codes of a visual nature.

The basic premise of a discursively oriented analysis of news values is that "news values can be seen as discursively constructed, and newsworthiness becomes a quality of texts. News values are thus defined as the 'newsworthy' aspects of actors, happenings and issues as constructed through discourse" (Bednarek & Caple, 2014: 137; Caple & Bednarek, 2017, pp. 3–4), and thus an analysis of the semiotic textual resources mentioned earlier, which (in the manner typical of semiotic research), focuses not on the process but on the outcome of this process in the form of a broadly defined text (syntagma), allows their specificity to be captured in terms of the news values encoded by these resources.

The DNVA is based on a modified catalogue of news values, which are described in detail in terms of both linguistic and visual communication elements (cf. Bednarek & Caple, 2012b, 2017; Caple & Bednarek, 2017). The elements relevant for the further analyses in this chapter are presented in Table 6.2.

Bednarek and Caple propose adopting two perspectives for analysing news values in the context of visual messages: content and camera techniques. Content takes into account what is depicted in an image at the level of denotation (Barthes, 1977), it includes the participants, their actions and the circumstances in which these actions take place. Camera techniques, on the other hand, take into account the way in which the image is captured (composition, the way information is arranged in the image frame, perspective, including angle; technical considerations, for example, shutter speed, aperture, etc.) and the way in which the image is captured (Caple & Bednarek, 2017). At the theoretical level, this division is justified, but generates some difficulties at the analytical level. Firstly, the analysis of camera techniques requires an in-depth technical knowledge of photography, but—much more importantly and as illustrated in Table 6.1—in practice, news values are encoded through the sum of the

Table 6.2 News values and visual codes

News value	*Characteristic*	*Visual codes*
Aesthetic Appeal	The event is discursively constructed as beautiful (visuals only)	Content: showing people, places, objects and landscapes culturally recognised for their beauty. Camera technique: dynamic, asymmetric composition, making use of diagonal axis; lighting, colour contrast and shutter speed used for artistic effect.
Consonance	The event is discursively constructed as (stereo) typical (limited here to news actors, social groups, organizations, or countries/ nations)	Content: showing happenings that fit with the stereotypical imagery of an event/person/country and so on (e.g. images of beer and breasts for the Oktoberfest).
Eliteness	he event is discursively constructed as of high status or fame (including but not limited to the people, countries, or institutions involved)	Content: showing known and easily recognisable key figures, for example, political leaders, celebrities; showing people in uniform or with other regalia of officialdom; showing self-reflexive elements like microphones/cameras, media scrum; showing people being flanked by military, police or bodyguards; showing context associated with an elite profession, for example, books, lab, police station. Camera technique: low camera angle indicating high status of participant in image.
Impact	The event is discursively constructed as having significant effects or consequences (not necessarily limited to impact on the target audience)	Content: showing the after-effects (often negative) of events, for example, scenes of destruction; showing emotions caused by an event; showing sequences of images that convey cause and effect relations.

(*continued*)

Table 6.2 (continued)

News value	*Characteristic*	*Visual codes*
Negativity	The event is discursively constructed as negative, for example, as a disaster, conflict, controversy, criminal act	Content: showing negative events and their effects, for example, the aftermath of accidents, natural disasters, the injured/ wounded, the wreckage/ damage done to property; showing people being arrested or (as defendant) with lawyers/barristers/police; showing people experiencing negative emotions. Camera technique: igh camera angle, putting viewer in dominant position (often used with photographs of minors or offenders/ prisoners of war); Ccamera movement and blurring, combined with camerapeople moving around, running, ducking to avoid projectiles and so on (suggesting unstable situation, that is, danger).
Personalization	The event is discursively constructed as having a personal or 'human' face (involving non-elite actors, including eyewitnesses)	Content: showing 'ordinary' individuals, especially when singled out and standing in for a larger group; showing an emotional response. Camera technique: positioning individuals in unequal relation to others in the image frame, for example, singling out one individual through foregrounding or backgrounding; using a close-up shot (to focus on a person's emotion, for example).
Positivity	The event is discursively constructed as positive, for example, as a scientific breakthrough or heroic act	Content: showing positive events and their effects, showing people experiencing positive emotions.

(*continued*)

Table 6.2 (continued)

News value	*Characteristic*	*Visual codes*
Proximity	The event is discursively constructed as geographically or culturally near (in relation to the publication location/target audience)	Content: showing well-known or iconic landmarks (Tower Bridge, Sydney Opera House, Golden Gate Bridge), natural features (Uluru) or cultural symbols (flags, national colours/distinctive uniforms); showing verbal text indicating relevant place/cultural connection, for example, signage.
Superlativeness	The event is discursively constructed as being of high intensity or large scope/scale	Content: showing the repetition of key elements in the image frame, for example, not just one house but an entire street affected; showing extreme emotions in participants; placement of elements of different sizes next to each other to maximise contrast, for example, the tallest and the shortest member of a group. Camera technique: use of specific lens (wide vs telephoto) and angle to exaggerate or condense differences in size/space; camera movement and blurring, combined with camerapeople moving around, running, ducking to avoid projectiles and so on (suggesting seriousness/high danger, etc.).
Timeliness	The event is discursively constructed as timely in relation to the publication date: as new, recent, ongoing, about to happen, or otherwise relevant to the immediate situation/time (current or seasonal)	Content: showing relevant time in the image, for example, the season may be implied in flora or environmental conditions; inclusion of cultural artefacts, like Christmas trees that are representative of a timely event; showing verbal text indicating relevant time, for example, signagei.
Unexpectedness (Novelty)	The event is discursively constructed as unexpected, for example, as unusual, strange, rare	Content: showing people being shocked/surprised; showing unusual happenings that would be considered outside an established societal norm or expectation; juxtaposition of elements in the frame that create stark contrast.

Source: Bednarek and Caple (2012), Caple and Bednarek (2017)

denotative and the connotative, and for some of them it is difficult to distinguish characteristic technical elements. For this reason, the analyses presented in this chapter ignores technical issues, treated as separately coded features of fake news.

DNVA understood in this way and implemented in research practice allows, on the one hand, an analysis of how news values are constructed depending on the type of medium and technology (which, as discussed above, condition the use of specific semiotic codes) and depending on the specific type of news discourse (e.g. hard news versus soft news). In the research presented here, these assumptions will be applied to the analysis of news values realised by fake news in social media. As the authors point out, DDNVA can identify common practices, conventions, and clichés of news reporting and offer insights into news as semiotic practice, either at a particular point in time or across news cycles. Moving beyond this micro level of semiotic construction, it is also possible to use this type of analysis to explore if particular topics-such as indigenous news actors, asylum seekers/refugees, marriage equality, or climate change-are associated with specific news values. Such repeated associations may then have ideological implications, and DNVA can thus be used as a tool for critical discourse analysis (for further discussion of the critical potential of DNVA, see: Bednarek & Caple, 2014). Again, it is possible to undertake such analysis diachronically and across cultures. The aim here is to see if specific news values are emphasised, rare, or absent in reporting on particular topics or events, and in how far this is constrained by the event itself (Bednarek & Caple, 2017).

6.3 Research: Conception, Problematics, Method

With reference to the theoretical assumptions presented above, a study was carried out which referred to various aspects of visual elements (mainly photographs) included in fake news in Poland, the Czech Republic and Slovakia.

The analysed fake news originated from social media, which means, that the role of visual messages is even more important in their case than in the case of hard news, which is published on the Internet. This is due to the specifics of social media (striving for increased emotionality, emphasis on attracting the attention of the recipient), but also to the structure of the messages themselves. As noted by Bednarek and Caple (2017: 173), in social media (both in terms of messages generated by media institutions and fake news) visuality is what dominates spatially, taking up much more space than text. Posts on social media platforms are very limited in terms of textual volume, while emoticons, links and precisely the visual element play a special role in them.

The overall aim of the research was, in line with the DNVA, to both describe and attempt to explain the visual phenomena that occur in fake news. The question, therefore, is not only which visual elements are characteristic of fake news, but also what their function is—both in relation to the linguistic text and in—reconstructing—the relationship with the viewer.

As in the case of text research, the assumption was to create a research concept combining elements of qualitative and quantitative analysis (from a medium level, see: Chap. 5), in line with the DNVA. Since the aim was to compare these phenomena in three countries—Poland, Slovakia and the Czechia—it was decided to forego an in-depth analysis of individual images (carried out, for example, within the framework of the visual grammar concept) in favour of collecting data from a larger corpus of messages and attempting to grasp the regularities present in these data (corpora).

The corpus consists of visual elements of fake news from January 2023–April 2024,[1] which were verified as such on the AFP fact-checking portal in Poland (https://sprawdzam.afp.com/), the Czech Republic (https://napravoumiru.afp.com/) and Slovakia (https://fakty.afp.com/). The corpus was built on the base of a total of 182 visuals ($N = 182$). Formally, the analysis was performed with reference to the fact-checker archive, which is linked from publications posted on AFP websites). A justification of the corpus structures and the method used to create it (along with an indication of its limitations) was provided in the previous chapter of this

[1] The time span of the corpus is related, on the one hand, to the completion date of the previous phase of the research (Stasiuk-Krajewska, 2024) and, on the other, to the planned publication date of the research results. In terms of content, it was considered that such a time frame would be adequate to the issue and purpose of the research, due to the continuous nature of disinformation activities (including the production and versification of fake news). In view of this, it is difficult to distinguish a time frame whose boundaries could be justified by specific events occurring in the social or political world. On the other hand, structuring the corpus in this way made it possible to capture a range of phenomena that provoke disinformation activities, such as armed conflicts, elections, or health issues, and to avoid inadequacies related to possible thematic dominants (such as in the COVID-19 period). AFP was chosen as a source of verified fake news because, as a global agency, it debunks false information in Poland as well as in the Czech Republic and Slovakia. This made it possible to obtain comparable (e.g., in terms of selection criteria) material for all three countries studied.

book. The visual corpus included 71 visual messages in Polish ($N_p = 71$), 37 visual messages in Czech ($N_{cz} = 37$) and 74 visual messages in Slovak ($N_s = 74$). It should be emphasised that each visual syntagma present in the message was analysed separately. This means that if a fake news text was accompanied by a series consisting of, for example, three pictures, all three pictures were analysed separately. On the other hand, if the visual element of the fake news was a collage (the visual elements were combined into a coherent semantic whole), this was treated as a single message. The corpus was manually annotated for categories deemed relevant based on the literature and the author's previous research (see below for details).

The main research questions were formulated as follows:

1. What is the dominant format of visual messages in fake news?
2. What are the denotations present in these messages (i.e.: what can be seen)?
3. What are functions of visual materials in relations to the overall fake news message?
4. Which news values are constructed in the fake news corpus by visual messages and to what extent is this construction similar to what we deal with in real news (social media)?

The data obtained in the context of all the research questions posed were analysed both holistically (for crossover purposes) and comparatively (for individual countries). Data coding was manual and based on assumptions that applied Bednarek and Caple (2017, see below).

6.4 Research: Presentation and Analysis of Results

At the first stage of the study, questions were asked about the format of the graphic materials analysed. Audiovisual materials were omitted, on the assumption that they build a different type of relationship with the viewer through their different structure. When analysing these materials, other theoretical and methodological assumptions should also be used, for example those referring to narrative or storytelling theory. Of course, given the high popularity of this type of message in fake news (see previous chapter), such research should be conducted (Chart 6.1).

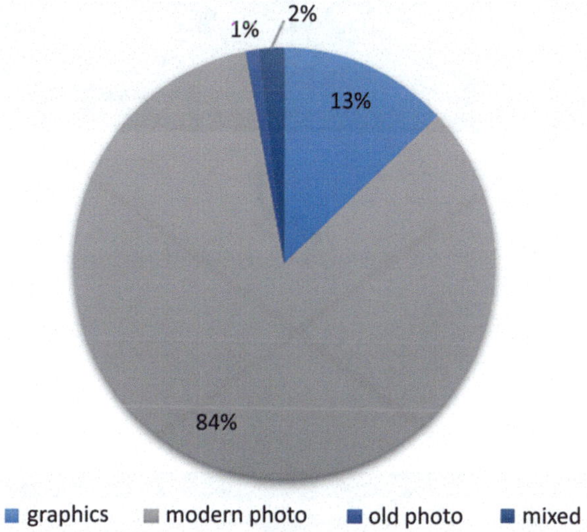

Chart 6.1 Format

As far as the format of static visual messages is concerned, the vast majority of these are photographs that, at least formally, are documentary in nature—they therefore present a 'true picture of reality' ('the reality principle', Stocchetti, 2017). In this respect, a similarity with journalistic photography can be observed, although the analysis of news values carried out at a later stage of the research revealed that the issue is not so obvious. It is interesting to note the relatively high presence of the 'graphics' category, which consists of texts presented in graphic form, sometimes fictitious print media satirical covers, but above all maps. Such representations are essentially hybrid in nature, carrying, for example, practically no classic news values (with the exception of satirical graphics). Their function is illustrative and documentary, and they often have the task of simply presenting the most important information (data). The map occupies a special position because, on the one hand, it conveys a lot of apparently reliable information, and on the other, it is associated with communication based on a high level of knowledge and competence (Fig. 6.1).

As far as the comparative aspect is concerned (Chart 6.2), a significant difference can be seen in Poland, where messages of a graphic nature (especially maps and texts) are more numerous. Old photographs are

Fig. 6.1 Map. (Source: https://sprawdzam.afp.com/doc.afp.com.33CT92A)

practically absent from fake news, and it is also rare to find a mixed format that combines graphics with reportage photos.

An interesting phenomenon in terms of format is the modifications that visual messages undergo (Chart 6.3).

As can be seen, ¼ of the messages analysed are modified. These modifications basically have two forms—it is either a collage, which creates a coherent syntagma from various elements, or a series, which juxtaposes several images. The phenomenon of the series is particularly interesting, as it has many important functions for fake news: it authenticates or intensifies the message, creates contrast or tells a story. In general, therefore, series, which are infrequent in hard news, have a clear persuasive function, directing the receipient of the message (Fig. 6.2).

The highest number of modified images is found in Slovakia and the lowest in the Czech Republic, with the difference being quite significant (Chart 6.4). It seems that the modification of images, while having a strong effect on the viewer, at the same time distances the visual material

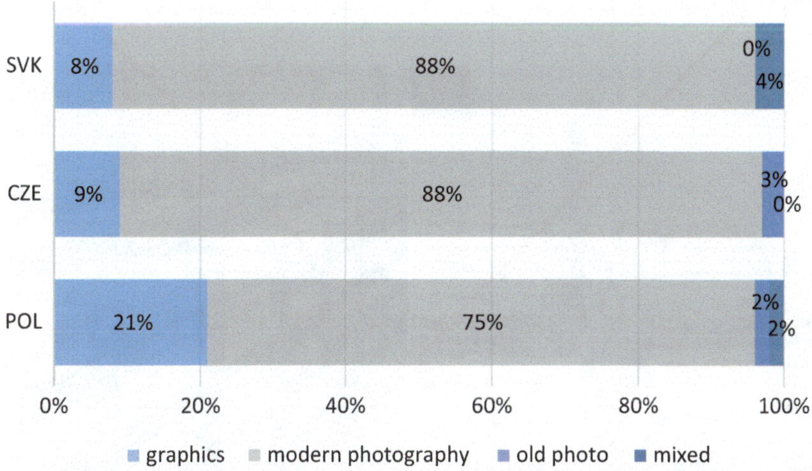

Chart 6.2 Format—comparison

Chart 6.3 Modifications

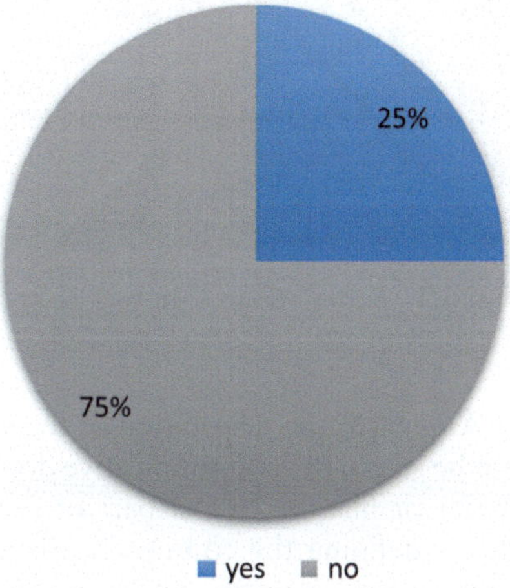

present in fake news from the journalistic standard. Under this assumption, the low number of such visual materials in Czech news would correlate with a tendency towards increased textual content—in the sense that it would make fake news in the Czech Republic more similar to real news.

Fig. 6.2 Series. (Source: https://fakty.afp.com/doc.afp.com.33778PK)

As regards the denotation of the visual messages in fake news (i.e. what can be seen), the communicative dominant was analysed, i.e. what is in the center or fills the largest part of the image (in the case of difficulties in clearly defining the communicative dominant, the message was classified in the varia category, which explains its relatively strong presence). In terms of denotation, the visual messages accompanying fake news are decisively dominated by images of people, which account for almost half of the denotation. It is therefore clear that fake news, like news, personalises—it shows the world through the prism of people, people who are positively or negatively evaluated. Representations of people also have an important communicative and functional advantage—namely, they can encode very many news values and perform diverse functions (e.g. sympathy for the

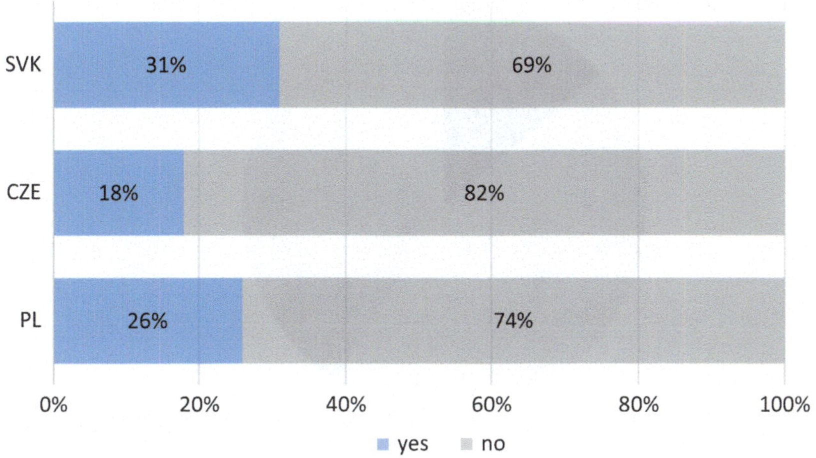

Chart 6.4 Modifications—comparison

victim, outrage at politicians, the truth about the behaviour of celebrities). The number of documents presented in photographs is also relatively high—the function of such messages is obviously to give credibility, to give something special meaning, to reveal what is covered. What is different are texts—these are graphics that do not present documents, but some data in a format prepared by the author. These graphics are usually more emotionally charged and less 'objective' than the presentation of documents (Chart 6.5).

As can be seen (Chart 6.6), fake news is the most personalised in Slovakia, while it is the least personalised in Poland, where, in turn, objects are relatively often depicted without people present. A denotative characteristics of visual messages in the Czech Republic is the strong position of the category 'documents', which again brings Czech fake news firmly into line with real news (Fig. 6.3).

A closer analysis of the way people are presented in the images that accompany fake news reveals two clear and interesting trends (Charts 6.7, 6.8, 6.9, 6.10, 6.11, and 6.12)

While the proportion between the portrayal of groups and the portrayal of individuals tends to be balanced (by the smallest degree in Slovakia), the pictures are definitely dominated by men, who appear in ¾ of the messages (it is worth noting that these are pictures showing only men, mixed

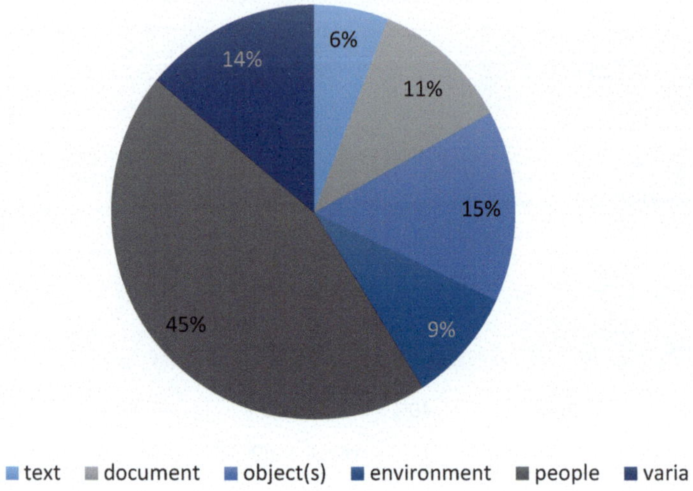

Chart 6.5 Denotation

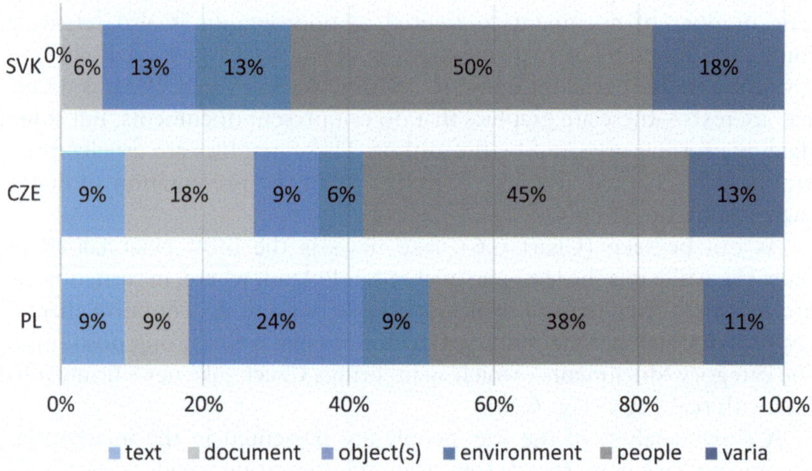

Chart 6.6 Denotation—comparison

groups are the 'other' category). This is probably related to the subject matter of fake news, which concerns (as mentioned in the previous chapter) issues related to politics, defence or conflict. In this way, fake news confirms

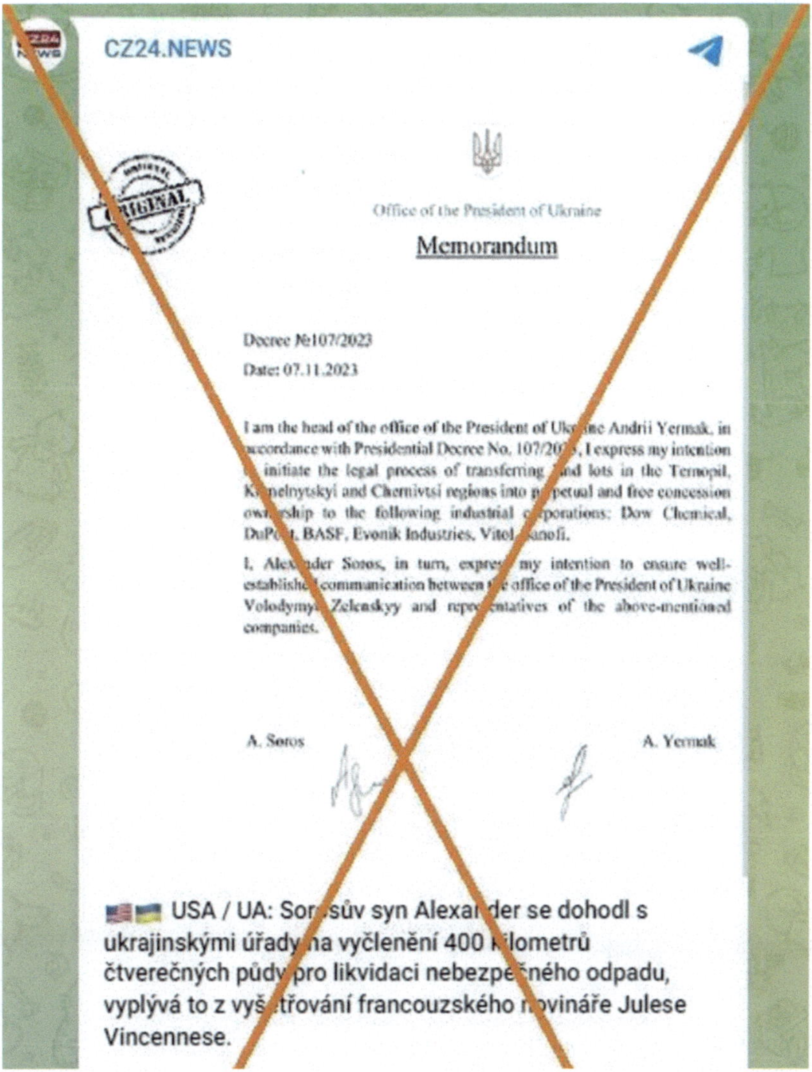

Fig. 6.3 Document. (Source: https://cedmohub.eu/cs/dajn-prezidentsk-dekrety-nedokazuj-e-ukrajina-jednala-se-sorosovm-synem-o-pevodu-pdy-na-zpadn-korporace/)

Chart 6.7 How many people?

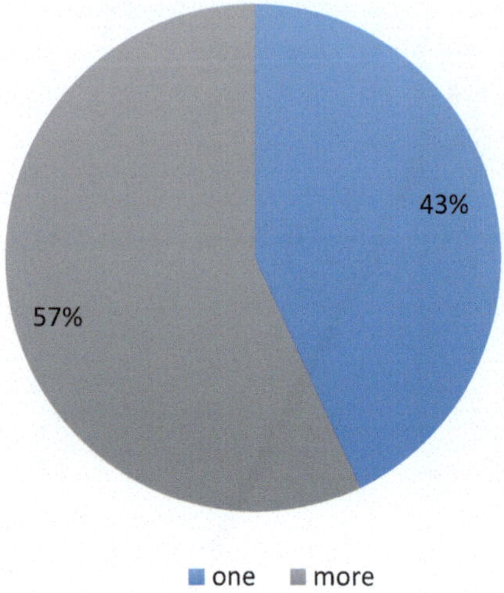

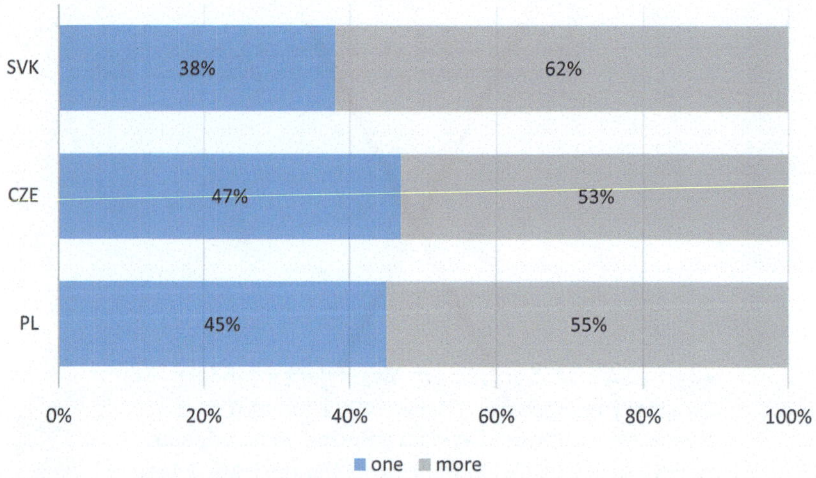

Chart 6.8 How many people—comparison

the vision of the world as male-dominated, and moreover sustains the discursive construct of masculinity—as associated with power, strength, courage, but also violent tendencies. Matters of importance are associated with the masculine rather than the feminine. The overwhelming dominance

Chart 6.9 Gender

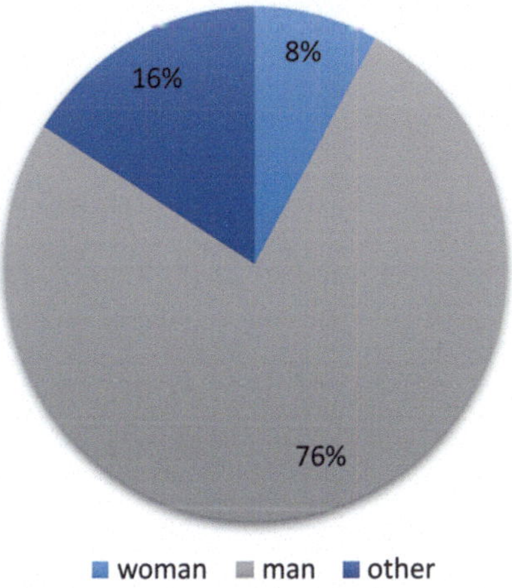

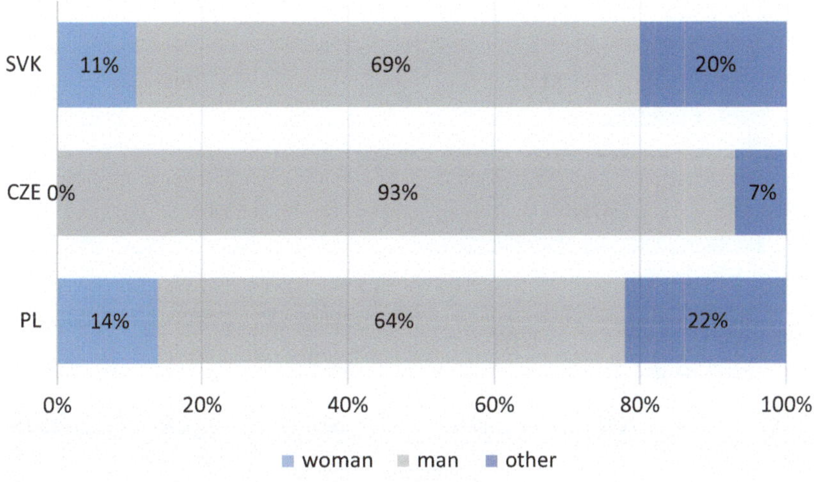

Chart 6.10 Gender—comparison

of men is observed in Czech messages, where (in the sample analysed) women are simply absent (Fig. 6.4).

The second characteristic is the very clear dominance of politicians (public officials). Their images account for almost 60% of all news coverage (most in Poland, least in Slovakia), which is incomparably more than

Chart 6.11 Roles

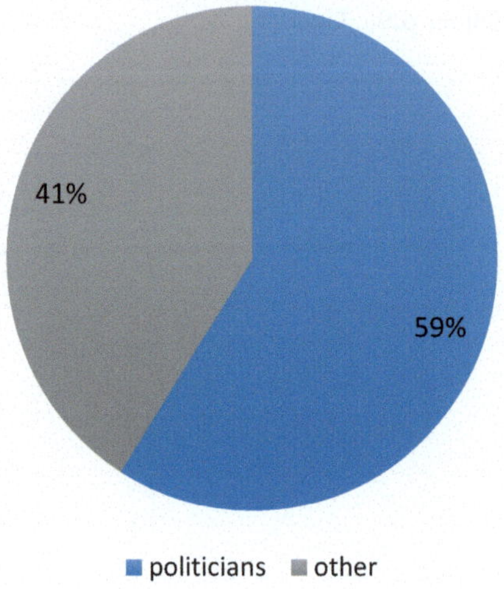

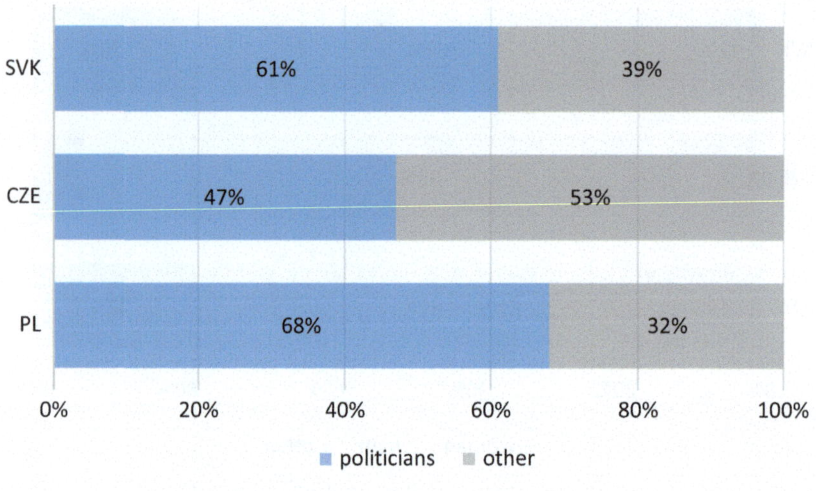

Chart 6.12 Roles—comparison

any other distinguishable professional group (soldiers, policemen or doctors also appear). This structure is again linked to the subject of fake news, but also constructs a very specific vision of the world, in which politics and politicians occupy a special place. It is a vision of the world similar to that

Fig. 6.4 Men. (Source: https://sprawdzam.afp.com/doc.afp.com.33U27MD)

of hard news. Of course, the context in which politicians are portrayed is overwhelmingly negative, and there are some favourite 'heroes' of fake news—first and foremost Volodymyr Zelenski, but also representatives of the European Union and the WHO. To a small extent, these are local politicians (Fig. 6.5).

The analysis of the functions of visual messages follows the classification proposed by Bednarek and Caple (2012). The authors list the following functions of news photography:

Fig. 6.5 Politican / Official. (Source: https://sprawdzam.afp.com/doc.afp.com.33FX86A)

1. Image as illustration. This is the role traditionally attributed to journalistic photography; images illustrate, complementing the textual message, which is considered the most important.
2. Image as evidence. Here, photography is recognised as the most reliable way of presenting the truth about the world that 'never lies'. In this function, the photograph is very clearly linked to the information function and the value of objectivity and factuality. This function is also most often attributed to journalistic photography in scientific analysis and journalistic training.
3. Image as sensation. Such images are mainly found in tabloids and are associated with coverage of scandals and the private lives of celebrities.

4. Image as icon. These types of images function as symbolic representations of key historical moments and are primarily associated with the depiction of wars, but also natural disasters or social tragedies.
5. Image as evaluation. This function is based on the ability of visual messages to convey and evoke strong emotions, both positive and negative. It involves a strong emotional impact on the recipient of the message.
6. Image as aesthetic. This function is controversial, just as the issues of photography as art are controversial. Even if it appears in press photography, it is not frequent.

The analysis of the function of visual messages in fake news showed that image as icon, image as aestetic, and image as sensation were not represented in the studied corpus (to some extent traces of image as sensation function could be found in photographs presenting evidence of certain behaviours, but this does not concern celebrities, only politicians and is not of a moral scandals nature). Due to the specificity of fake news, the image as evidence category was divided into two subcategories. The first of these ("evidence") includes visual messages that construct the main content of the information, show what is to be revealed. In this case, the text has only a supporting role and could not exist without the image. In the second case ("additional evidence"), the pictures support the text, add something to it, reinforce the message, present visual evidence. In the case of the illustrative function, the photos remain connected to the text, but have only an illustrative function (sometimes they are loosely connected to the textual message), for example showing a neutral photo of a politician in a official situation (Chart 6.13).

As can clearly be seen, the photo in the evidence function clearly dominates the others (the "evidence" and "additional evidence" categories total more than 50% of the messages). If we include the illustrative function we get more than 80% of the corpus. This phenomenon is interesting because, as can be seen, the visual messages in fake news have a relatively small emotional function, focusing rather on the factual function, supporting fake news in building the impression that it is presenting the true version of the world, revealing what is covered up. Thus, they bring fake news closer to how real news is defined in journalistic discourse. The graphic elements in fake news also have no aesthetic ambition, no ambition to represent the world metaphorically. Their role is, we will repeat, essentially informative (Fig. 6.6).

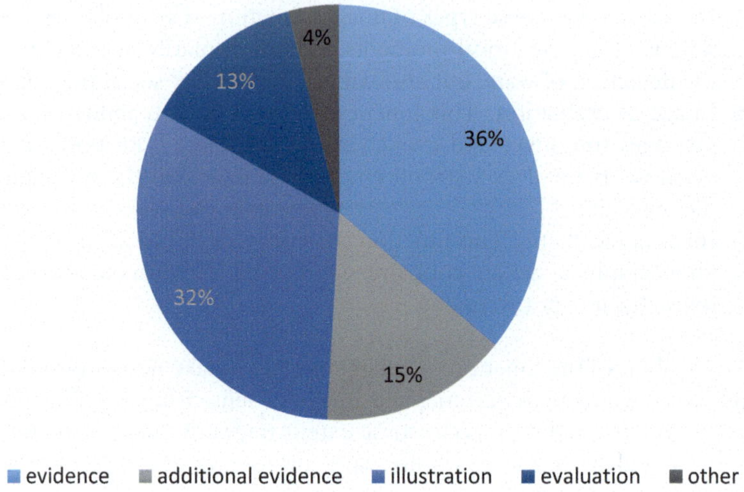

Chart 6.13 Function

When it comes to comparing the function of the graphic material, there are no major differences between countries. In each case, the observation regarding the dominance of the information function is confirmed (Chart 6.14).

The idea and implementation of the study on visual valuesof fake news follows the research carried out by Caple and Bednarek (2017). The authors analysed 1100 images captured from Facebook posts from 22 news organisations, the results of their study are presented in Table 6.3. The data obtained in Bednarek and Caple's study provide an excellent reference point for the study of fake news, as they concern real news in the social media space. They are therefore based on this corpus, which should provide a comparative context for the study of fake news.

The sample on which the research presented below was carried out is much smaller than in the study on news values in real news, so it should be regarded as a pilot study.

Obviously, in most of the messages analysed, more than one new value can be found, which was reflected in the coding method. Individual visual messages were assigned between 0 and 4 news values (which was the maximum number observed). Messages that were not assigned news values were twofold. These were, similar to the Bednarek and Caple study, stock

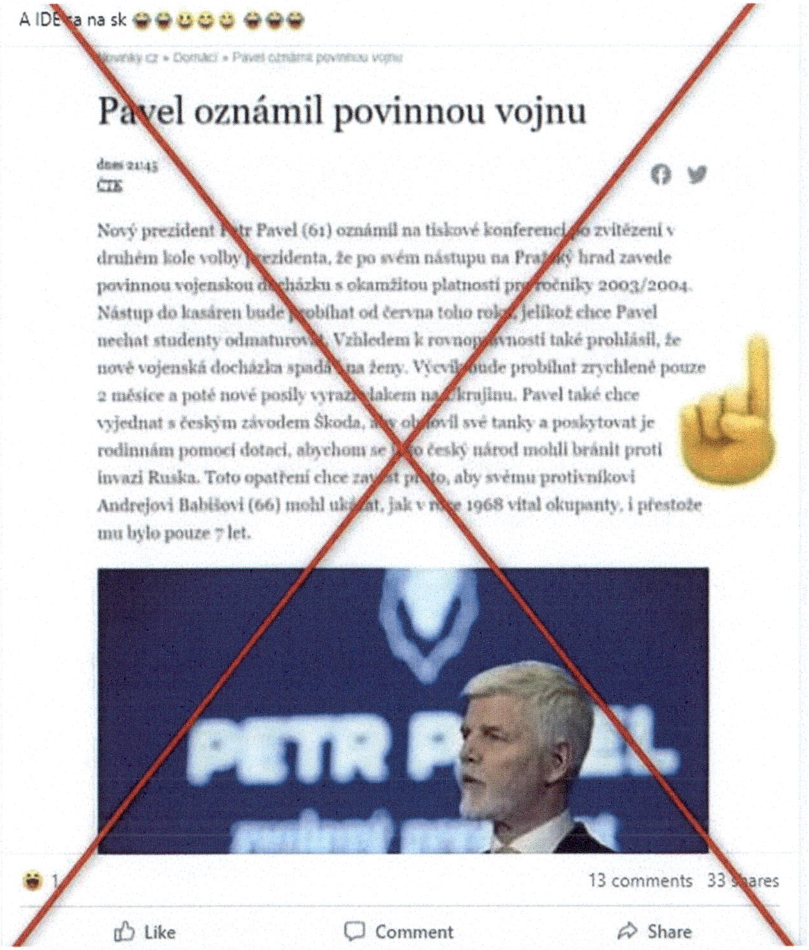

Fig. 6.6 Image as illustration. (Source: https://fakty.afp.com/doc.afp.com.33888HT)

photos. This type of photo is increasingly common in the media, although rather in the context of specific classes of media and texts and rather soft news. As the authors note, these types of photos do not depict any specific moments, but rather refer to general classes of people and phenomena (Bednarek & Caple, 2017: 91). The second category of visual messages

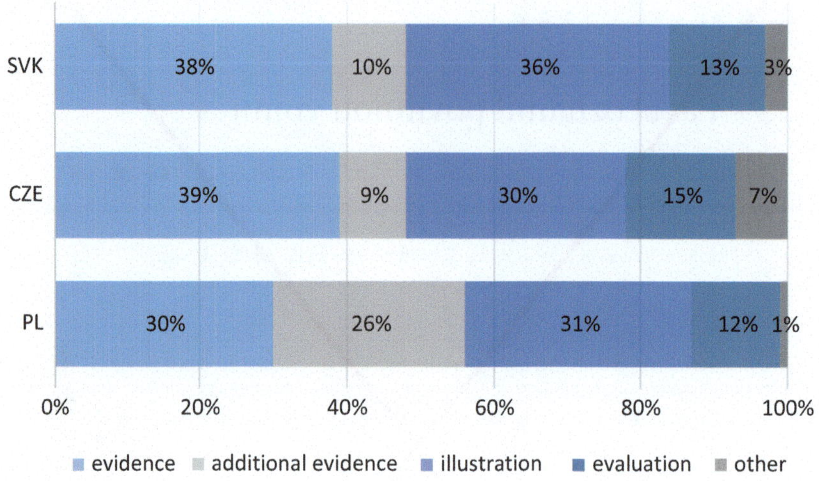

Chart 6.14 Function—comparison

Table 6.3 Visual news values in real news

News value	%
Elitness	33
Personalization	29
Proximity	29
Negativity	26
Positivity	19
Unexpectedness	12
Superlativeness	10
Impact	8
Aesthetic Appeal	2
Consonance	9
Timeliness	0

Source: Bednarek and Caple (2017)

that were not attributed news values in the study of fake news was specific to this type of message—namely photos of documents. The percent of messages that did not represent any news values in the real news corpus was 8%, in the fake news corpus it was located at a similar level (9%). At the same time, the vast majority were document photos (only 4 stock photos were found in the entire corpus). Thus, it can be said that the share of stock photos is relatively small in the fake news, which is directly related to the fact that they resemble hard news (and stock photos are the domain of soft news). In addition, if we assume that the predominant function of visuality in the case of fake news is that of presenting evidence, the uselessness of stock photos, which cannot exist in this function, becomes obvious.

Examples of the coding of individual news values in fake news are presented in Figs. 6.7, 6.8, 6.9, 6.10, 6.11, 6.12, 6.13, 6.14, 6.15, 6.16, 6.17, and 6.18.

Fig. 6.7 Elitness. (Source: https://cedmohub.eu/cs/jemen-nevyhlsil-izraeli-vlku/)

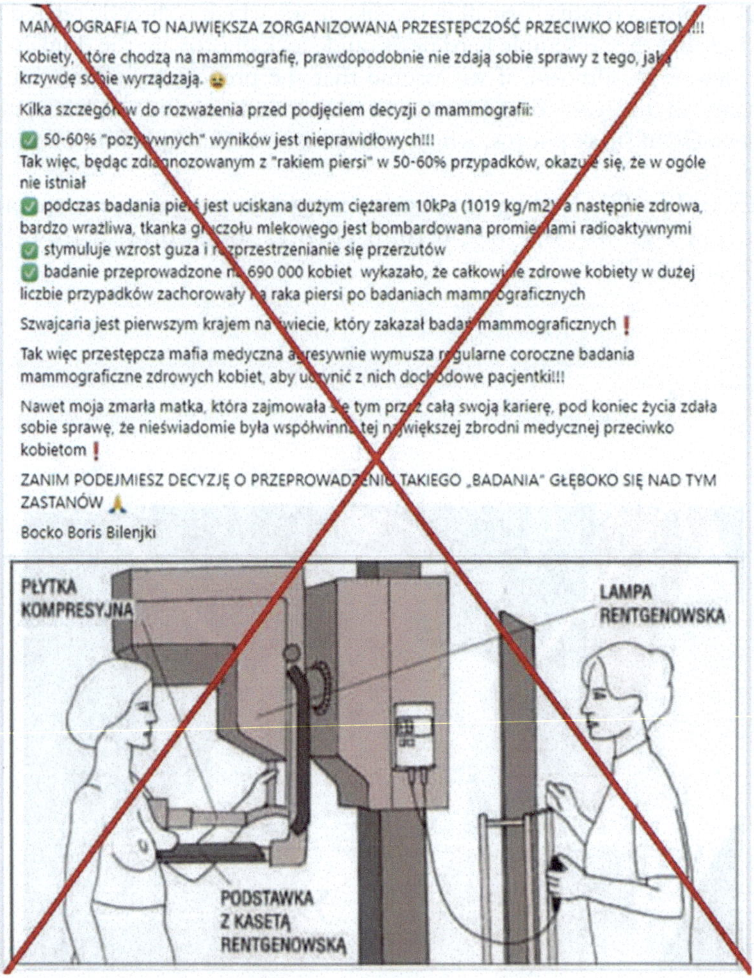

Fig. 6.8 Personalisation. (Source: https://sprawdzam.afp.com/doc.afp.com.3472277)

The percentage distribution of news values encoded by the visual elements of fake news is presented in Table 6.4.

As can clearly be seen, elitness definitely dominates the analysed messages. Elitness is mainly associated with the portrayal of politicians or

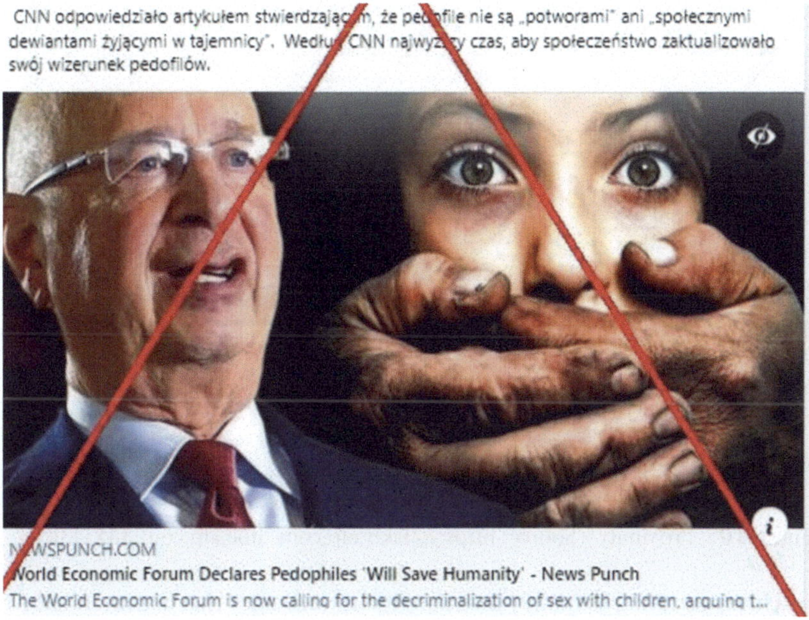

Fig. 6.9 Negativity. (Source: https://sprawdzam.afp.com/doc.afp.com.337H3BD)

public officials, possibly their families, and clearly correlates with negativity (in Poland, this correlation is 37%; in the Czech Republic 53%; in Slovakia 39%).

Negativity, the second most important news value, is (besides politicians and officials) encoded by tragic and violent phenomena (volcanic eruption, fires), situations with negative connotations (the role of negative protagonist is mainly played by soldiers or police officers), graphic elements superimposed on images (e.g. a prohibition or radioactivity sign), harm to ordinary people or images of the victims of the system (to a limited extent, as we will discuss in a moment).

Proximity, the third most recurring news value, is realised primarily through the presentation of local politicians. They are recognisable because they are the central figures of the photography, but also because they appear in contexts that support their recognisability (e.g. with the logo or name of the party they represent).

Fig. 6.10 Proximity. (Source: https://fakty.afp.com/doc.afp.com.33XU3KV)

Unexpectidness is mainly associated with messages that are intended to arouse outrage or disgust, possibly curiosity, but often in the context of a sense of fear or threat (such as in the case of the giant skull or the 'rain of worms').

Consonance refers to national stereotypes (Ukrainians, Russians) and stereotypes of socially excluded groups. These types of messages are decidedly negative.

Superlativeness refers primarily to a large number of something (which is sensational or frightening, repellent) and is achieved, in large part, through the use of a series of images (e.g. worms or gold).

Values that lie below 10% are: aesthetic appeal, impact, positivity, personalisation and timeliness. Aesthetic appeal relates predominantly to landscapes, to a lesser extent to works of human hands. In a small proportion of cases, visual messages are subjected to aestheticisation processes (e.g. through colour intensification), but this should be considered an exceptional situation. Impact, as in the case of real news, involves the presentation of negative consequences of events, but does not occur frequently. Similarly, personalisation, which, let us recall, refers to showing 'ordinary people' (so depictions of politicians are not included in this category). The category of positivity is interesting. These are images that,

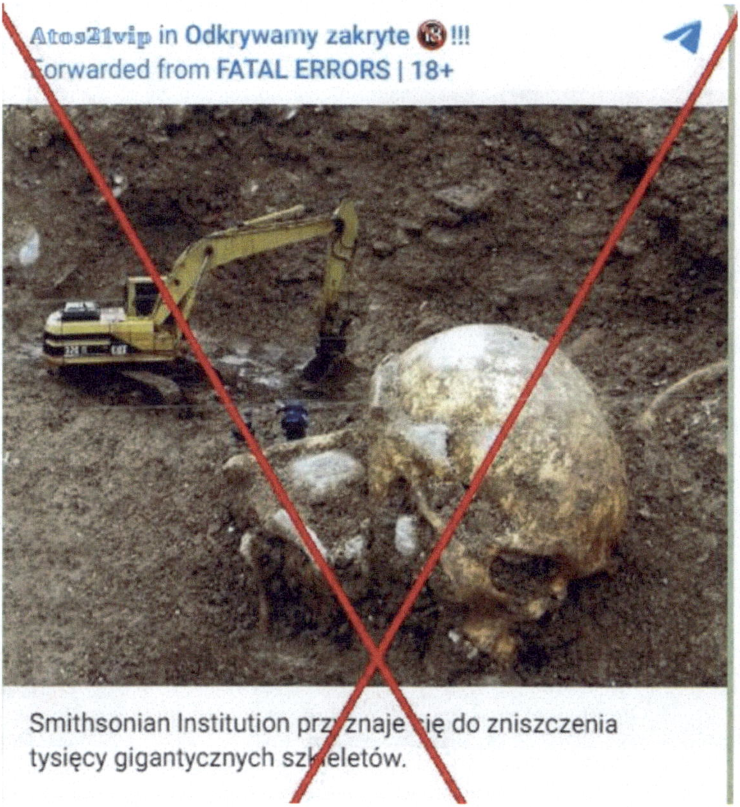

Fig. 6.11 Unexpectidness. (Source: https://sprawdzam.afp.com/doc.afp.com.33BR4RV)

when analysed in isolation from the text, encode positive emotions (e.g. through the facial expressions of the people), but in the context of the text take on other connotations, most often ironic or negative.

The analysis also revealed interesting differences between countries (Table 6.5)

Although the set of the first two news values (elitness and negativity) and their mutual correlation remain constant, interest in (critical) power elites is much higher in Slovakia than in Poland and the Czech Republic. In Poland, messages are more stereotypical (consonance) than in the

Fig. 6.12 Consonance. (Source: https://sprawdzam.afp.com/doc.afp.com.339P8TJ

Czech Republic and Slovakia, and there is a significant predominance of references to aesthetics (especially to 'Polish landscapes', for example in the context of grain imports from Ukraine). Proximity, which can be interpreted as an expression of interest in local issues (mainly local politics), is in turn much lower in Poland than in Slovakia and especially in the Czech Republic. In Slovakia, the graphic elements of fake news also seem to indicate that it is slightly more sensational (unexpectidness). In all cases, personalisation, timeliness and impact rank low.

Chiny. Spadł "deszcz robaków" w Pekinie. Władze apelują o użycie parasoli

dzisiaj, 09:46 | Świat

Fig. 6.13 Superlativeness. (Source: https://sprawdzam.afp.com/doc.afp.com.33F44E7)

A comparison of the news values connoted by the visual messages in fake news with the visuals of real news also shows significant and interesting differences (Table 6.6).

Admittedly, eliteness still comes first, but its share is much higher in fake news than in real news. Negativity is also much higher, while positivity is lower. Unexpectendess is slightly higher, which is linked to the evocation of negative emotions and to the presentation of images that could raise

Fig. 6.14 Aestetics. (Source: https://sprawdzam.afp.com/doc.afp.com.33ER743)

questions in terms of journalistic ethics. Consonance and superlativeness are also higher. This indicates a desire to attract, to astonish the viewer and a clear construction of a schematic vision of the world based on stereotypes and prejudices. Most characteristic is a very significant difference a in

Fig. 6.15 Impact. (Source: https://fakty.afp.com/doc.afp.com.33TT8PF)

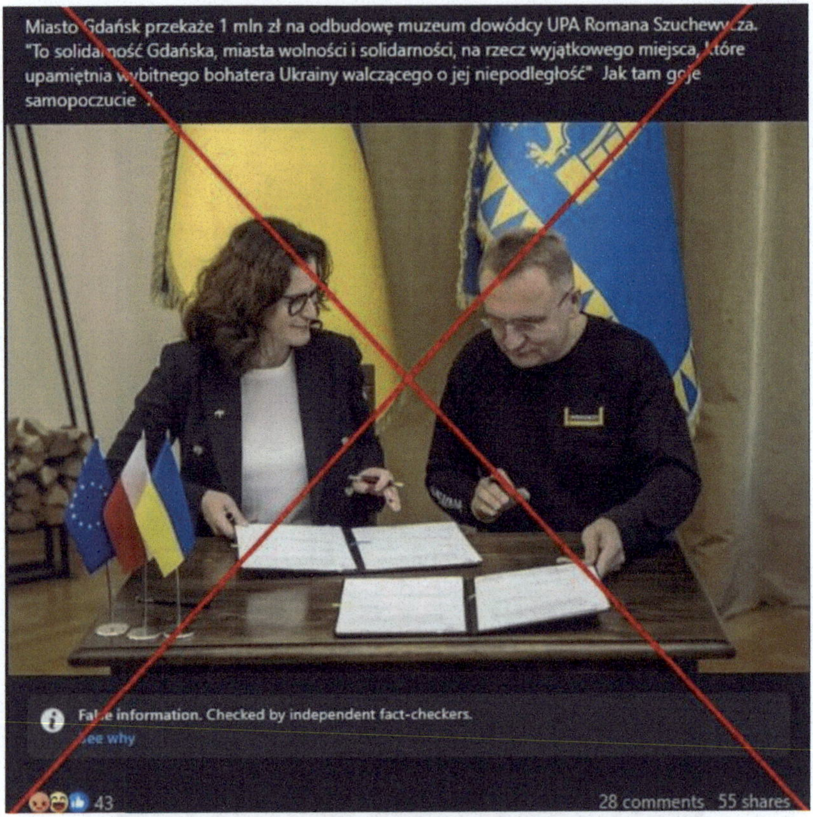

Fig. 6.16 Positivity. (Source: https://sprawdzam.afp.com/doc.afp.com.34GB8TG)

terms of personalisation. Fake news, surprisingly, is not interested in average people, their lives, their tragedies or their successes.

The differences observed above also lead to a change in the most common clustering. While in real news elitness is linked to positivity or proximity, in fake news it is linked to negativity. Significant clustering is not created by personalisation because its level, in general, is very low.

Research on news values coded by real news has also been carried out by Harcup and O'Neill (2017). However, the authors refer to other categories, which makes comparison difficult. For example, in their studies,

Fig. 6.17 Timeliness. (Source: https://sprawdzam.afp.com/doc.afp.com.33DJ9ZQ)

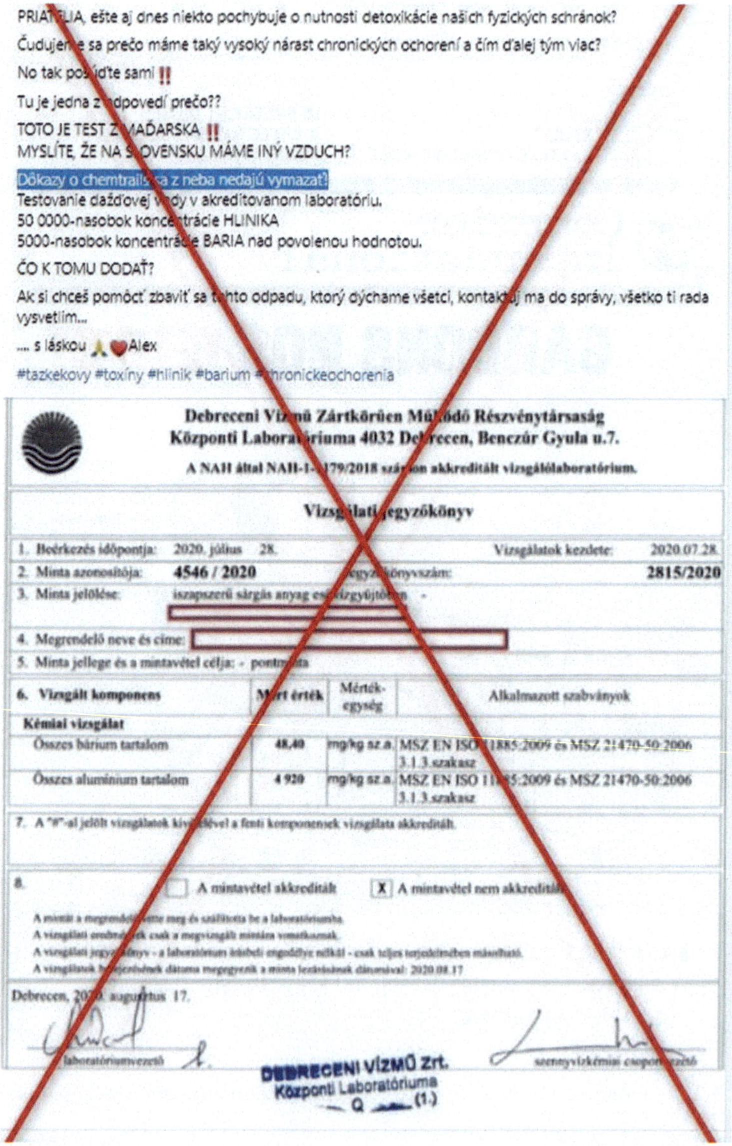

Fig. 6.18 No news values. (Source: https://cedmohub.eu/cs/pekarov-adamov-nenavrhla-financovat-zdravotnickou-pi-pro-ukrajinsk-uprchlky-privatizac-budvaru/)

Table 6.4 New values in fake news

News value	%
Eliteness	55
Negativity	43
Proximity	26
Unexpectedness	22
Consonance	20
Superlativeness	17
Aesthetic Appeal	8
Impact	9
Positivity	9
Personalization	6
Timeliness	2

Table 6.5 News values in fake news—comparison

Poland	%	*Czechia*	%	*Slovakia*	%
Eliteness	46	Eliteness	48	Eliteness	70
Negativity	42	Negativity	48	Negativity	38
Consonance	25	Proximity	32	Unexpectedness	37
Aesthetic appeal	17	Consonance	16	Proximity	28
Proximity	17	Unexpectedness	16	Superlativeness	26
Unexpectedness	14	Superlativeness	13	Consonance	20
Impact	13	Impact	10	Positivity	11
Superlativeness	13	Positivity	10	Personalization	6
Positivity	7	Personalization	6	Impact	5
Personalization	6	Aesthetic appeal	3	Aesthetic appeal	3
Timeliness	1	Timeliness	3	Timeliness	1

the category of bad news comes first, but let us note that it is only apparently relevant to the category of negativity. Bad news as such (e.g. about natural disasters) is relatively scarce in fake news, but critical attitudes towards the world predominate, which should be classified as negativity. Next in line are surprise and entetrainment, which would also be difficult to transfer directly to a fake news analysis. Note that these two categories also correlate in the study in question with the popularity of stories published on social media. As it seems, in the case of fake news, the mechanism of this popularity is definitely different (based on criticism of authority and a negative, stereotypical vision of the world), although this would obviously require further research.

Table 6.6 News values in fake news and real news—comparison

Fake news	%	*News*	%
Eliteness	55	Elitness	33
Negativity	43	Personalization	29
Proximity	26	Proximity	29
Unexpectedness	22	Negativity	26
Consonance	20	Positivity	19
Superlativeness	17	Unexpectedness	12
Aesthetic appeal	8	Superlativeness	10
Impact	9	Impact	8
Positivity	9	Aesthetic appeal	2
Personalization	6	Consonance	9
Timeliness	2	Timeliness	0

6.5 Summary and Conclusions

The above data indicate that the most characteristic features of visual representations in fake news include:

1. In terms of format: predominance of reportage-like images; modifications, especially in the form of series.
2. In terms of denotation: the dominance of representations of persons, mainly men and politicians.
3. In terms of functions: the dominance of functions related to documenting, showing evidence.
4. In terms of news values: strong dominance of eliteness and negativity, very low level of personalisation.

As far as the visuality of fake news in Poland is concerned, it is distinguished by a higher proportion of graphic representations, a relatively low (compared to other countries) number of representations of people, more stereotyping and a relatively low interest in local issues.

In the Czech Republic, visual messages are the least modified, present a relatively large number of documents and are characterised by a high degree of localness.

In Slovakia, we observe the highest percentage of message modification and the highest representation of individuals. Also the highest critical interest in the authorities and slightly more sensationalism than in the other countries studied.

Data on the format, denotation and function of visuals in fake news indicate that visual messages reinforce the similarity of fake news to real news. They operate mainly with realistic images in the function of describing the world and presenting evidence, avoiding stock photos. A specific feature is the modifications that appear in ¼ of the analysed material. These are mainly series that are not typical of real news (although, let us note, they are already beginning to appear there, for example in the context of sports news). The analysed materials present mainly people, thus implementing assumptions related to personalisation (although, as indicated by the analysis of news values, understood differently than in real news).

The data on the values encoded by visual messages in fake news, especially when compared to those in real news, reveal a specific construction of the world in fake news. It is decidedly a male world, perceived negatively and critically, as a source of threats and unauthorised actions by a representative of power. In general, 'men of power' are what visual representations of fake news are most interested in. Thus, power is seen as something that creates order in the world and, at the same time, something decidedly negative. But this negativism is mainly personal—power is bad because the politicians who represent it are bad. Thus, it is not a negation of power and politics as such, but of specific people associated with it (in the analysed corpus—primarily Zelensky). Some of the people in power are local politicians, whose representations constitute the category of proximity. The world constructed by the visual elements of fake news is therefore not the world of ordinary people and their everyday life (surprisingly low position of the personalisation value). Indeed, people are shown in photographs, but they are people of power, the social elite. They are the focus of fake news. If "ordinary people" appear, it is as those wronged by power, but note that even such depictions are few. Fake news also exploits the classic news values that can be seen in real news: sensationalism, stereotyping or influence.

The research presented in this chapter certainly calls for continuation. Above all, it would be worthwhile to repeat them on a larger corpus, and to link the observed forms, denotations and news values to the popularity of fake news in social media and thus to include the perspective of the recipient. A further extension of the geographical scope of the research would also certainly have a significant impact on the representativeness of the findings for fake news as a global phenomenon.

References

Allan, S. (Ed.). (2010). *The Routledge companion to news and journalism*. Routledge.

Barthes, R. (1977). *Image, music, text*. Fontana.

Bateman, J. A. (2011). The decomposability of semiotic modes. In K. L. O'Halloran & A. S. Bradley (Eds.), *Multimodal studies: Multiple approaches and domains*. Routledge.

Bednarek, M. (2006). Epistemological positioning and evidentiality in English news discourse: A text-driven approach. *Text & Talk, 26*(6), 635–660.

Bednarek, M. & Caple, H. (2012a). 'Value added': Language, image and news values Discourse. *Context & Media, 1*(2–3), 103–113.

Bednarek, M., & Caple, H. (2012b). *News discourse*. Bloomsbury.

Bednarek, M., & Caple, H. (2014). Why do news values matter? Towards a new methodological framework for analysing news discourse in critical discourse analysis and beyond. *Discourse & Society, 25*(2), 135–158.

Bednarek, M., & Caple, H. (2017). *The discourse of news values: How news organizations create newsworthiness*. Oxford University Press.

Bell, A. (1991). *The language of news media*. Blackwell.

Bennett, W. L. (2016). *News: The politics of illusion*. University of Chicago Press.

Caple, H., & Bednarek, M. (2013). *Delving into the discourse: Approaches to news values in journalism studies and beyond*. Working Paper. The Reuters Institute for the Study of Journalism, The University of Oxford.

Caple, H., & Bednarek, M. (2017). Rethinking news values: What a discursive approach can tell us about the construction of news discourse and news photography. *Journalism, 17*(4), 435–455.

Chouliaraki, L. (Ed.). (2012). *Self-mediation: New media, citizenship and civil selves*. Routledge.

Cotter, C. (2010). *News talk: Investigating the language of journalism*. Cambridge University Press.

Deuze, M. (2005). What is journalism? Professional identity and ideology of journalists reconsidered. *Journalism, 6*(4), 442–464.

Fowler, R. (2013). *Language in the news: Discourse and ideology in the press*. Routledge.

Galtung, J., & Ruge, M. H. (1965). The structure of foreign news: The presentation of the Congo, Cuba and Cyprus crises in four Norwegian newspapers. *Journal of Peace Research, 2*(1), 64–90.

Golan, G. J. (2008). Where in the world is Africa? Predicting coverage of Africa by US television networks. *International Communication Gazette, 70*(1), 41–57.

Hall, S. (1973). The determinations of news photographs. In S. Cohen & J. Young (Eds.), *The manufacture of news: Deviance, social problems and the mass media* (pp. 176–190). Constable.

Hall, S., Critcher, C., Jefferson, T., Clarke, J., & Roberts, B. (1978). *Policing the crisis: Mugging, the state and law and order*. Macmillan Press.
Harcup, T., & O'Neill, D. (2001). What is news? Galtung and Ruge revisited. *Journalism Studies, 2*(2), 261–280.
Harcup, T., & O'Neill, D. (2017). What is news? News values revisited (again). *Journalism Studies, 18*(12), 1470–1488.
Harrison, J. (2006). *News*. Routledge.
Hartley, J. (1996). *Popular reality: Journalism, modernity, popular culture*. Arnold.
Hartley, J. (2013). *Understanding news*. Routledge.
Hartmann, P., & Husband, H. (1976). The mass media and racial conflict. In S. Cohen & J. Young (Eds.), *The manufacture of news: Social problems, deviance and the mass media*. Constable.
Herman, E. S., & Chomsky, N. (2010). *Manufacturing consent: The political economy of the mass media*. Vintage Digital.
Hodge, B., & Kress, G. (1998). *Social semiotics*. Polity Press.
Kress, G. (2010). *Multimodality: A Social Semiotic Approach to Contemporary Communication*. Routledge.
Kress, G., & van Leeuwen, T. (1996/2006). *Reading images: The grammar of visual design*. Routledge.
Kress, G., & van Leeuwen, T. (2001). *Multimodal discourse*. Hodder Arnold.
Kress, G., & van Leeuwen, T. (2002). Colour as a semiotic mode: Notes for a grammar of colour. *Visual Communication, 1*(3), 343–368.
Laclau, E., & Mouffe, C. (1985). *Hegemony and socialist strategy: Towards a radical democratic politics*. Verso.
Lemke, J. L. (1999). *Typology, topology, topography: Genre semantics*. MS University of Michigan.
Lippmann, W. (2020). *Liberty and the news*. MediaStudies Press.
Manghani, S., Piper, A., & Simons, J. (2006). *Images. A reader*. Sage.
Meinhof, U. H. (1994). Double talk in news broadcasts: A cross-cultural comparison of pictures and texts in television news. In D. Graddol & O. Boyd-Barrett (Eds.), *Media texts: Authors and readers* (pp. 212–223). Open University Press.
Mirzoeff, N. (2002). The subject of visual culture. In M. Mirzoeff (Ed.), *The visual culture reader*. Routledge.
Palczewski, M. (2018). *Teorie newsa*. Elipsa.
Patterson, T. E. (2013). *Informing the news: The need for knowledge-based journalism*. Vintage Books.
Schlesinger, P. (1978). *Putting 'reality' together: BBC news*. Constable.
Schudson, M. (2003). *The sociology of news*. W.W. Norton & Company.
Schulz, W. F. (1982). News structure and people's awareness of political events. *International Communication Gazette, 30*(3), 139–153.
Soloski, J. (1989). News reporting and professionalism: Some constraints on the reporting of the news. *Media, Culture & Society, 11*(2), 207–228.

Stasiuk-Krajewska, K. (2024). Elementy wizualne fake newsów. In K. Stasiuk-Krajewska & M. Wenzel (Eds.), *Struktury i funkcje. Dezinformacja w czasach kryzysu*. Adam Marszałek.
Stocchetti, M. (2017). Digital visuality and social representation: Research notes on the visual construction of meaning. *KOME-An International Journal of Pure Communication Inquiry, 5*(2), 38–56.
Tuchman, G. (1972). Objectivity as strategic ritual: An examination of newsmen's notions of objectivity. *American Journal of Sociology, 77*(4), 660–679.
Tuchman, G. (1978). *Making news: A study in the construction of reality*. Free Press.
van Dijk, T. A. (1988a). *News as discourse*. Routledge.
van Dijk, T. (1988b). *News analysis: Case studies of international and national news in the press*. Lawrence Erlbaum Associates.
van Leeuwen, T. (1999). *Speech, music, sound*. Macmillan Press.
van Leeuwen, T. (2005). *Introducing social semiotics*. Routledge.
Wahl-Jorgensen, K., & Hanitzsch, T. (Eds.). (2020). *The handbook of journalism studies* (2nd ed.). Routledge.
Zelizer, B. (2004). Language studies and journalism. In B. Zelizer (Ed.), *Taking journalism seriously: News and the academy*. Sage Publications.
Zelizer, B. (2010). *About to die: How news images move the public*. Oxford University Press.

CHAPTER 7

The Carry-over Effect in a Polycrisis: How Attitudes to Covid-19 Mutated into Vulnerability to Russian Disinformation During the War in Ukraine

Michał Wenzel

The recent global crisis situations, including the Covid-19 pandemic and the war in Ukraine, have sometimes been called a "polycrisis" (Lawrence et al., 2024). Such a designation stresses the causal chains between individual crises: they are not isolated occurrences, but reinforce one another. This chapter starts with the assumption that the social reactions to both of these disruptions were linked. Chronologically, the pandemic preceded the war in Ukraine by 2 years. The restrictions on social life during Covid-19 were unprecedented and involved testing the limits of human rights (Spadaro, 2020). Restrictions were contested in many countries and some policies were rejected by significant groups in society. State-enforced limitations on freedoms elicited attitudes which, in our view, remained durable. We hypothesize that the attitude to the subsequent crisis, i.e. the

M. Wenzel (✉)
Dept. of Social Sciences, SWPS University, Warsaw, Poland
e-mail: mwenzel@swps.edu.pl

M. Wenzel, K. Stasiuk-Krajewska (eds.), *Disinformation in Central Europe*, https://doi.org/10.1007/978-3-031-76871-2_7

Russian invasion of Ukraine, was a consequence of the generalized attitudinal rejection of the political consensus; a sceptical attitude formed during the pandemic was embedded in society and caused the negation of the consensus around the war in Ukraine. Thus, it increased people's susceptibility to Russian disinformation tactics.

Our research was conducted in Poland, the Czech Republic and Slovakia. These three countries were affected by both stages of the polycrisis. We used a range of statistical methods to determine which pockets of society were particularly susceptible to disinformation activities in terms of socio-demographic attributes and media consumption. We assumed that: a. There would be differences between the countries related to historical legacies and state policies, and b. Vulnerability would be related to how citizens obtain information about world events. The analysis is based on findings from mass surveys representative of adult populations.

7.1 Narratives About Covid-10 and the War in Ukraine

In the chapter "Strategies of Disinformation in Central Europe: Narratives Related to the Covid-19 Pandemic and War in Ukraine" in this volume, the modus operandi of various state and non-state actors in the case of the C-19 pandemic and the war in Ukraine is discussed. It is noted that both traditional and alternative media were active in promoting disinformation narratives about Covid-19, for the benefit of a number of stakeholders. The disinformation narratives concerned the existence of the disease, virus origins and the effectiveness of vaccines, as well as other aspects of the pandemic.

The war in Ukraine closely followed Covid-19, and in the information ecosystem it was marked by a heightened level of uncertainty and distrust in the elites. The Covid-19 narratives in traditional media changed over time, and many aspects of the pandemic were only gradually proved or disproved (e.g. the effectiveness and side effects of the vaccines), while other aspects have still not been conclusively clarified (e.g., the origins of the virus). These circumstances caused fertile ground for the spread of unverified information. This was not necessarily disinformation: our definition (see Chap. 2) treats disinformation as spreading falsehood with evil intent, i.e. inferring that an actor is behind it. In the case of C-19 narratives, there was both deliberate disinformation and a spontaneous social process of knowledge creation.

In the case of both the Covid-19 crisis and the war in Ukraine, Russian state-affiliated actors are documented as being the creators and disseminators of disinformation narratives (Hoyle et al., 2022; Milewski, 2020; Doroshenko & Lukito, 2021; Khaldarova & Pantti, 2020). Moreover, there is some evidence (Rzeczkowski et al., 2024; Kovacheva, 2023) that individuals, organizations and media that questioned the official narratives about C-19 moved on to question the western and NATO policy towards the war in Ukraine. They may have been agents of disinformation, or representatives of groups and individuals who were excluded from public discourse.

In our research we pose a thesis that this mechanism operated not only on the level of media actors or public organizations, but on the level of public opinion as well: scepticism towards the official state policy regarding C-19 hypothetically translated into a negative attitude towards policy regarding the war, and hence a pro-Russian attitude. An attitude (the concept is understood as defined by Allport, 1929) developed, whereby a generalized scepticism towards the traditional media resulted in a rejection of the framing of the official narrative in both cases. In both the C-19 crisis and the war in Ukraine, the political elites, both the government and most opposition parties, developed a policy consensus reflected in the main media outlets. Some people questioned this consensus and looked for an alternative narrative. Those who reject the dominant narrative may suspect a collusion of powerful actors (Terracciano, 2023).

7.2 Media Environment in Central Europe

Mass media in Central and Eastern Europe, including the Visegrad countries, followed the political trajectory of these states: from the state-controlled and censored media under the pre-1989 Communist regime, to an open media system with both state and private outlets under the democratic regime within the legislative framework of the European Union (Dobek-Ostrowska & Glowacki, 2015). The changes were marked by a number of pressures and challenges. Among them, three are the most important for the analysis of the interplay between media use and disinformation.

First, there is the difficult interrelation between the media and politics. This is partially a legacy of the former system (Jakubowicz, 2001). The negotiated transition prevented a radical overhaul of the media and politics. The public (state) media went through a prolonged period of transition, which allowed managers and journalists associated with the autocratic regime to continue their modus operandi after transition. Subsequently, public media were the spoils of parties after gaining power, which resulted in a politicised editorial line following the government policy (Bajomi-Lázár, 2014; Dobek-Ostrowska, 2019). Additionally, private media developed ties with political parties and civil society linked to political groups. This resulted in so-called "political parallelism", i.e. a media system in which partisan media represent the worldviews and interests of political groupings (Hallin & Mancini, 2004; De Albuquerque, 2018). Such a system undermines both the impartiality of the media and the quality of the political system, in particular its accountability.

The second challenge to media quality and impartiality are commercial entanglements. Newspapers, magazines, radio and TV stations are owned by private companies and individuals. In some cases these are multinational corporations, such as Ringiar Axel Springer. In other cases they are people pursuing business interests in several sectors, such as Zygmunt Solorz-Żak in Poland or Zdenek Bakala and Daniel Kretinski in the Czech Republic (Dębiec, 2021). There exists a conflict of interests in such media when it comes to reporting about the business activities of its owners. In some cases, media have been owned by active politicians. This is the case of Andrej Babis, the former Czech Prime Minister. In this case, the conflict of interests concerns political reporting.

The political and commercial pressures are unequally distributed in the countries under study. In Poland, party parallelism and political control have been a cause of concern. The takeover by the government of the state media and the resulting bias in reporting contributed to the decline of Poland in the rankings of free speech (Freedom House, 2024). Moreover, the political polarization reflected in media alignments was replicated on the level of the audience: the political cleavages covered political parties, the media and citizens, who were divided according to their political views in their media use (Wenzel, 2018).

Commercial pressure was strong in the Czech Republic and Slovakia. The process of nationalisation or bohemization of the media occurred in the 2000s, whereby the media formerly owned by big multinational corporations changed hands and became the property of Czech and Slovak tycoons. Business practices with regard to the media arms of their operations were called into question (Dębiec, 2021; Stetka, 2012; Dragomir, 2020).

The third challenge to the media system is global: it is the rise of social media as a source of information. There are multiple threats inherent in the substitution of traditional media (in both print, electronic and online versions) with social media. Among the most egregious are: the de-professionalization of journalism, the potential for disinformation due to either lack of verification of the news or through deliberate distortion, and the tabloidization of the news (infotainment as the main source of information) (Crilley & Gillespie, 2019; Wiik, 2019). Moreover, information freedom may be impaired by the editorial policies of the owners of social media platforms (Stjernfelt & Lauritzen, 2020).

7.3 Hypotheses and the Model of the Analysis

Taking into account our assumptions and previous research on media systems in Central Europe, we have arrived at a number of hypotheses about the patterns of media consumption and their relation to the disinformation narratives concerning the Covid-19 pandemic and the war in Ukraine.

The first hypothesis is about the structuring of media use patterns with regard to information about the public sphere. "Rather than choosing between a few available traditional media outlets, people nowadays select a combination of media platforms to consume news. This selection can include both online and offline platforms from both institutionalized and non-institutionalized news media. Due to this societal and technological change, a more comprehensive approach is warranted in measuring news consumption by examining news media repertoires" (Geers, 2020: 333–334). We expected the audience to gather information according to three patterns. The first pattern relates to using state (public) media. Public media in Central Europe form a distinct ecosystem (Dobek-Ostrowska, 2019). Reading the major private newspapers or magazines and tuning in to private radio and TV channels falls outside of its bounds, forming a second pattern. The third pattern involves using social media to learn about the world. Social media users form a distinct group of information-seekers (Geers, 2020). The repertoires overlap and we expected blurred boundaries and regularities in use, rather than exclusive behaviour.

While there is significant overlap in the use of different outlets, some media are more trusted than others. Some content is verified elsewhere, while some is not. We believe, as our hypothesis number two, that using social media as the most trusted (i.e. not double-checked) source of information may lead to users being especially vulnerable to disinformation narratives (Shu et al., 2017). This stems from the unprofessional character of the news, as well as from the potential to infiltrate them by state and non-state actors. Both possibilities must be taken seriously: there may be deliberate activity to introduce material, or a bottom-up process without external intervention. Moreover, the boundaries of disinformation may be blurred and they may shift, as happened in case of Covid-19. What is important about social media for our hypothesis is their autonomy vis-à-vis the political and media elites.

Thirdly, we expected differences between the three societies under study. Differences in attitudes may be especially pronounced in relation to the war in Ukraine. The attitude towards Russia and state policy regarding the war, in particular the hawkish stance in Poland and weak support for Ukraine in Slovakia may be reflected in public attitudes.

Fourthly, we put forward the carry-over hypothesis. We believed that the attitude to Covid-19 vaccines was a reflection a latent anti-elite sentiment and is a predictor of the scepticism to the official narrative regarding the war in Ukraine. There is some evidence to suggest that there was a "rally around the flag" effect in Europe during both crises (or polycrisis) (Nicoli et al., 2024). We are putting forward a similar argument about the people who refused to rally around the flag: that developing an opposition to the state policy in the first crisis (C-19 pandemic) morphed into a lasting attitude.

The research plan was divided into three stages. First, we detected the patterns of media use. We enquired about broad categories and individual outlets as sources of information. We created indices using factor analysis to detect latent variables recording the patterns of media use.

Secondly, we constructed indicators of disinformation. We operationalized it as alternative narratives about the reality. We used CEDMO and EDMO research, as well as expert factcheckers in order to select dominant narratives, present in all three countries for comparison. We pilot-tested the items and narrowed down the list. Subsequently, we created synthetic indicators. The details are in Sect. 7.5.

Finally, we performed bivariate and multivariate analyses. The purpose was to detect the net influence of media use on susceptibility to disinformation. We controlled for standard socio-demographic factors.

The surveys were fielded in Poland, the Czech Republic and Slovakia on samples representative of the adult population (18+). The study was conducted using the CAWI (computer-assisted web interview) methodology. Quota sample was used in the Czech Republic and Slovakia, and random sampling in Poland. The fieldwork dates were: in Poland, June 20–28, 2022 (N = 1021); in the Czech Republic, June 1–17, 2022 (N= 1002); in Slovakia, June 17, 2022 (N= 1013). In the Czech Republic and Slovakia, the survey was carried out by the company STEM / MARK, a. S.; in Poland, by PBS.

7.4 Patterns of Media Use

The consumption of media, in our approach, is hypothetically a significant predictor of attitudes to the Covid-19 pandemic and the war in Ukraine. Mass media, however, may be used in different ways by categories of users. Broadly, a medium may be considered as fully credible and its content may be accepted without further verification. Alternatively, the medium may be only conditionally accepted. In such a case, its content will be verified in other, more credible sources.

The hierarchy of media categories in Poland is dominated by private news media, i.e. TV, radio stations and internet websites. Family members and friends also feature prominently as sources of knowledge about the world. Public radio and TV are much less trusted than private outlets. Social media are used by the vast majority of people in Poland, but are consumed without verification by a minority of users.

Czech users differ from Poles in that websites and public television are by far the most important sources of news for them. Public television and radio are much more important in the Czech Republic than private outlets. This contrasts with Poland, where public TV and radio are relatively seldom used. Friends and family members, like in Poland and Slovakia, are trusted by Czechs to give valid news.

Slovaks resemble Czech society in their hierarchy of news sources. The most important for them are internet websites and public television. In both the Czech Republic and in Slovakia, the use of social media is different to that in Poland. While in the former two societies most users trust them enough not to verify, in Poland most social media users check the news content in other sources (Table 7.1).

In all three countries, the media landscape is rich and diverse. Broad categories, such as print media, comprise tabloids as well as broadsheets.

Table 7.1 Sources of information, in categories

	PL		*CZ*		*SK*	
What are your sources of information?	*Used without verification*	*Verified elsewhere*	*Used without verification*	*Verified elsewhere*	*Used without verification*	*Verified elsewhere*
Private television (e.g. Polsat, TVN)	52%	32%	41%	18%	35%	22%
Websites	45%	50%	67%	35%	49%	34%
Family members	43%	45 %	48%	36%	43%	38%
Private radio stations	37%	27%	22%	11%	29%	21%
Friends	35%	53%	52%	37%	42%	38%
E-mail messages	33%	47%	24%	18%	22%	23%
Public radio stations	33%	25%	36%	19%	35%	25%
Social media	32%	50%	42%	22%	39%	27%
Public television	27%	27%	59%	24%	47%	28%
School and other educational institutions	26%	24%	21%	17%	26%	22%
Newspapers and magazines	22%	28%	27%	12%	25%	13%

In this region, political parallelism plays an important role: news outlets are tied to political parties and are frequently highly partisan. Therefore, the incumbency of a party in government and its links to a medium may affect the audience. In order to establish consumption patterns, it is important to determine the structure of audiences. As the first step in determining how audiences in the countries under study are structured, we asked questions about the sources of news: where respondents obtain information about public events. Multiple responses were possible.

In Poland, the most popular electronic media are the private TV channels TVN and Polsat, as well as the private radio stations RMF FM and Radio Zet (Table 7.2). Compared with these, public TV and radio are far less important as news sources. At the time of the survey, public (state)

Table 7.2 Sources of information, outlets[a]

Please indicate all sources where you obtain information	*PL*	*CZ*		*SK*	
		Television			
TVN, TVN24	60%	Česká televize	59%	Jednotka	42%
Telewizja Polsat, Polsat News	50%	FTV Prima	45%	TV Markíza	40%
TVP (public)	30%	Nova Group	42%	TV JOJ	40%
TV Republika	5%	Seznam.cz TV	30%	TA3	38%
Telewizja Trwam	4%	Barrandov TS	11%	Dvojka	23%
				TV Lux	7%
		Radio			
RMF FM	54%	Český Rozhlas	27%	Rádio Expres	34%
Radio Zet	40%	Rádio Impuls	20%	Rádio Slovensko	30%
Polskie Radio (publiczne)	15%	Frekvence 1	18%	Fun rádio	20%
Radio TOK FM	11%	Evropa 2	17%	Rádio Vlna	18%
Radio Maryja	4%	Rádio Blaník	12%	Rádio Európa 2	17%
		Rádio Kiss	11%	Rádio Jemné	16%
		Hitrádio	10%	Rádio Regina	11%
		Rock Rádio	7%	Rádio Patria	5%
		Rádio Beat	7%		
		Fajn Radio	6%		
		Country rádio	6%		
		Rádio Čas	4%		
		Newspapers and magazines			
Gazeta Wyborcza	26%	MF Dnes	14%	Sme	14%
Fakt	22%	Deník	12%	Pravda	13%
Super Express	18%	Lidové noviny	10%	Nový čas	12%
Local or regional dailies	14%	Hospodářské noviny	8%	Plus 7 dní	12%
Tygodnik Newsweek	13%	Metro	8%	Hospodárske noviny	11%
Rzeczpospolita	11%	Právo	8%	Denník N	11%
Dziennik Gazeta Prawna	11%	Reflex	8%	Zdravie	10%
Tygodnik Polityka	8%	Sport	8%	Život	8%
Gazeta Polska	7%	Blesk	7%	Plus 1 deň	8%
Tygodnik wSieci	5%	Respekt	6%	Nový čas pre ženy	7%
Nasz Dziennik	5%	E15	5%	Báječná žena	7%
Tygodnik Do Rzeczy	4%	5plus2	5%	Korzár	6%
		Téma	4%	Slovenka	6%

(*continued*)

Table 7.2 (continued)

Please indicate all sources where you obtain information	*PL*		*CZ*		*SK*
		Ekonom	4%	Új Szó	3%
		Forbes	3%		
		Internet portals			
Portal "Onet.pl"	60%	seznamzpravy.cz	53%	aktuality.sk	40%
Portal "wp.pl"	52%	novinky.cz	51%	topky.sk	26%
Portal "Gazeta.pl"	26%	iDnes.cz	45%	tvnoviny.sk	25%
Portal "Niezależna.pl"	8%	aktuálně.cz	39%	sme.sk	24%
Portal "naTemat.pl"	7%	ceskatelevize.cz / CT24	38%	pravda.sk	23%
		iprima.cz / CNN Prima News	27%	hnonline.sk	20%
		Deník.cz	24%	dennikn.sk	19%
		irozhlas.cz	14%	webnoviny.sk	18%
		lidovky.cz	13%	cas.sk	17%
		reflex.cz	11%	pluska.sk	16%
		Blesk.cz	9%	refresher.sk	15%
		forum24.cz	9%	dnes24.sk	15%
		echo24.cz	9%	dobrenoviny.sk	12%
		ceskenoviny.cz	5%	etrend.sk	8%
		drbna.cz	4%	teraz.sk	7%
		Social media			
Facebook	79%	Facebook	44%	Facebook	43%
YouTube	61%	YouTube	32%	YouTube	33%
Instagram	36%	Instagram	20%	Instagram	22%
TikTok	22%	TikTok	5%	TikTok	9%
Twitter (X)	21%	Twitter	9%	Twitter	8%
Snapchat	10%			Snapchat	4%
LinkedIn	8%	LinkedIn	5%	LinkedIn	4%

[a]The Czech and Slovak data are not directly comparable to the Polish. In the Czech Rep. and Slovakia respondents indicated whether they used a source without verification or verified it elsewhere. In the table are the percentages of people who use the source without checking the content elsewhere. In Poland respondents marked all sources used

media were controlled by the government of the right-wing Law and Justice party (PiS), and their coverage was highly sympathetic to the government and hostile to the opposition (Połońska, 2019). The most widely used internet portals Onet.pl and wp.pl are both used by the majority of respondents. Onet is owned by the international media corporation

Ringier Axel Springer while wp.pl is a publicly traded company controlled by Polish businesspeople. The major print publications are marked by political identity. Gazeta Wyborcza, Polityka, and Newsweek are left-liberal, with an editorial line supporting the opposition at the time of research (since 2023 in government), while Gazeta Polska, wSieci, Do Rzeczy and Nasz Dziennik are tied to PiS, in government during the survey fieldwork and since 2023 in opposition. Fakt and Super Express are apolitical tabloids. Among international social media platforms, the most popular are Facebook and YouTube, followed by Instagram, TikTok and X. If Facebook were to be treated as an independent source of information, it would be the most important one in Poland.

The Czech media landscape is in some respects different from the Polish example. The most common electronic sources of news (Table 7.2) are state radio and television (Česká Televise, Český Rozhlas). This is consistent with the data on the degree of implied trust in the media, as evidenced by accepting the content without verifying (Table 7.1). Moreover, internet portals related to Czech state media play a significant role in providing information. The role of commercial TV as news source is smaller than that of state TV. The most used print media include: MF Dnes and Lidové noviny, controlled then by the former Czech prime minister Andrej Babiš[1] (later sold). At the time of research he was in opposition to the government. He also controlled the radio station Rádio Impuls, as well as two important internet news sources lidovky.cz and iDnes.cz. The other title in the top three of print items is Denik, controlled by a Czech businessman.[2] As regards global social media, Facebook and YouTube occupy the top places.

The perceived reliability of Slovak electronic media sources resembles the situation in the Czech Republic, though the position of the state media is slightly weaker. State TV is more often used as a news source than private TV (Table 7.1). Consequently, the main public TV channel tops the ranking of TV stations, closely followed by the private outlets TV Markíza and TV JOJ (Table 7.2). As regards radio stations, the commercial Rádio Express is slightly ahead of Rádio Slovensko. Among the most important media sources is the website aktuality.sk. The hierarchy

[1] https://www.swissinfo.ch/eng/czech-billionaire-babis-s-media-assets-sold-ahead-of-new-law/48779426 [acccess 11.06.2024].

[2] https://www.osw.waw.pl/en/publikacje/osw-commentary/2021-02-11/bohemianisation-media-acquisition-czech-media-sector-local [acccess 11.06.2024].

of social media is similar to that in Poland and the Czech Republic, with Facebook, YouTube and Instagram as the most important sources.

The data about individual outlets is used in order to determine patterns of use. We extracted latent variables corresponding to regularities. Following our theoretical considerations, we expected people to use the media as news sources according to three patterns: relying on the state (public) media, main private outlets (radio and TV) or use of social media. In order to verify this hypothesis, we performed a confirmatory factor analysis. W extracted a three-factor solution. The analysis was performed separately for all three countries. The results are shown in Tables 7.3, 7.4 and 7.5.

The three-component factor analysis for Poland follows the hypothetical assumptions. Factor 1 groups the state and right-wing media: public radio and TV, as well as pro-government outlets: Nasz Dziennik, Gazeta Polska, the weeklies wSieci and Do Rzeczy, Radio Maryja and the internet portals associated with them. Factor 2 is the latent variable indicating the use of major private news outlets, print, traditional electronic media and internet, excluding the private outlets linked to the former PiS government. It includes the main broadsheets (Gazeta Wyborcza, Rzeczpospolita, Dziennik Gazeta Prawna), tabloids (Fakt, Super Express), opinion weeklies (Newsweek and Polityka), major TV stations (TVN and Polsat), biggest private radio stations (RMF FM and Radio Zet), as well as the largest internet platforms (Onet.pl and wp.pl). Factor 3 groups international social media, i.e. Facebook, YouTube, Instagram, TikTok, Snapchat and X (Twitter).

In the Czech Republic, the pattern of use is somewhat different, in that print media are distinct. Factor 1 groups all newspapers and magazines: Blesk, MF Dnes, Deník, Metro, Lidové noviny, Hospodářské noviny, Sport, Aha!, E15, 5plus2, Reflex, Téma, Respekt, Forbes, Týdeník Květy and Ekonom. Factor 2 is the latent variable indicating the use of most internet platforms (with the exception of Blesk.cz), as well as other selected media, such as the Czech public radio. Factor 3 includes the major social media platforms, but also most TV and radio stations, both private and public.

In Slovakia, as in Czech Republic, the use of print media (Hospodárske noviny, Nový čas, Plus 1 deň, Pravda, Sme, Denník N, Korzár, Új Szó, Plus 7 dní, Báječná žena, Život, Nový čas pre ženy, Slovenka, Zdravie), combined with the use of public radio stations (Rádio Slovensko and Rádio Patria) forms a distinct pattern (Factor 2 in Table 7.5). Factor 1 in

Table 7.3 Factor analysis. Outlets in Poland

Rotated component matrix			
PL		Component	
	1 (state/right wing media)	2 (major private outlets)	3 (social media)
Gazeta Wyborcza		0.618	
Rzeczpospolita		0.451	
Dziennik Gazeta Prawna		0.417	
Nasz Dziennik	0.499		
Gazeta Polska	0.489		
Fakt		0.383	
Super Express		0.410	
Local or regional dailies		0.385	
Tygodnik Polityka		0.386	
Tygodnik wSieci	0.539		
Tygodnik Do Rzeczy	0.582		
Tygodnik Newsweek		0.554	
TVP (public)	0.453		
TVN, TVN24		0.563	
Telewizja Polsat, Polsat News		0.492	
Telewizja Trwam	0.655		
TV Republika	0.470		
Polskie Radio (public)	0.556		
RMF FM		0.467	
Radio Zet		0.472	
Radio Maryja	0.565		
Radio TOK FM		0.369	
Portal "Onet.pl"		0.502	
Portal "wp.pl"		0.397	
Portal "Niezależna.pl"	0.477		
Portal "naTemat.pl"	0.407		
Facebook			0.532
YouTube			0.544
Instagram			0.713
TikTok			0.661
Snapchat			0.627
Twitter (X)			0.482

Extraction Method: Principal Component Analysis. Rotation Method: Varimax with Kaiser Normalization. Loadings above 0.3

Table 7.4 Factor analysis. Outlets in the Czech Republic

Rotated component matrix	*Component*		
CZ	1 (print)	2 (internet)	3 (electronic/social media)
Blesk	0.516		
MF Dnes	0.642		
Deník	0.607		
Metro	0.582		
Lidové noviny	0.614		
Hospodářské noviny	0.645		
Sport	0.570		
Aha!	0.454		
E15	0.555		
5plus2	0.473		
Reflex	0.631	0.300	
Téma	0.402		
Respekt	0.592		
Forbes	0.441		
Týdeník Květy	0.308		
Ekonom	0.536		
Česká televize			0.396
Nova Group			0.601
FTV Prima			0.593
Barrandov TS			0.373
Seznam.cz TV		0.328	0.365
Český Rozhlas		0.341	
Rádio Impuls			0.578
Rádio Blaník			0.518
Evropa 2			0.578
Frekvence 1			0.521
Hitrádio			0.445
Rádio Kiss			0.532
Rock Rádio			0.359
seznamzpravy.cz		0.618	
novinky.cz		0.603	
aktuálně.cz		0.644	
Deník.cz		0.561	
iDnes.cz		0.563	
iprima.cz / CNN Prima News		0.439	0.355
Blesk.cz			0.322
irozhlas.cz		0.567	
ceskatelevize.cz / CT24		0.575	

(*continued*)

Table 7.4 (continued)

Rotated component matrix	*Component*	
reflex.cz	0.508	
ceskenoviny.cz	0.408	
forum24.cz	0.516	
echo24.cz	0.506	
lidovky.cz	0.563	
Facebook		0.495
YouTube		0.460
Instagram		0.409
Pinterest		0.305
TikTok		0.310

Extraction Method: Principal Component Analysis. Rotation Method: Varimax with Kaiser Normalization. Loadings above 0.3

Table 7.5 Factor analysis. Outlets in Slovakia

Rotated component matrix		*Component*	
SK	1 (electronic media/ internet)	2 (print)	3 (social media)
Hospodárske noviny		0.562	
Nový čas		0.639	
Plus 1 deň		0.563	
Pravda		0.607	
Sme		0.593	
Denník N		0.534	
Korzár		0.546	
Új Szó		0.411	
Plus 7 dní		0.674	
Báječná žena		0.587	
Život		0.644	
Nový čas pre ženy		0.559	
Slovenka		0.620	
Zdravie		0.569	
TV Markíza	0.645		
TV JOJ	0.605		
Jednotka	0.548		
Dvojka	0.445		
TA3	0.484		
Rádio Slovensko	0.440	0.331	
Rádio Patria		0.304	0.448

(*continued*)

Table 7.5 (continued)

Rotated component matrix		*Component*	
Fun rádio	0.560		
Rádio Expres	0.564		
Rádio Vlna	0.382		
Rádio Európa 2	0.483		
Rádio Jemné	0.378		
pluska.sk	0.589		
topky.sk	0.609		
dennikn.sk	0.441		
dobrenoviny.sk	0.446		
dnes24.sk	0.476		
webnoviny.sk	0.590		
refresher.sk	0.484		
pravda.sk	0.548		
sme.sk	0.577		
cas.sk	0.518		
teraz.sk	0.327		
hnonline.sk	0.468		
tvnoviny.sk	0.653		
Facebook	0.430		0.360
YouTube	0.478		0.425
Instagram	0.367		0.498
Pinterest			0.451
TikTok			0.587
Snapchat			0.601
Twitter			0.574
LinkedIn			0.511

Extraction Method: Principal Component Analysis. Rotation Method: Varimax with Kaiser Normalization. Loadings above 0.3

Slovakia groups electronic outlets (TV and radio), as well as websites, in addition to some social media platforms. All social media load on Factor 3.

Overall, the patterns in the Czech Republic and Slovakia bear more similarity. Unlike in Poland, in these two countries the political orientation of the outlets has little influence on the use of media to obtain information. In both the Czech Republic and Slovakia, the use of print media is strongly correlated. In Poland, however, the editorial policy divides the readers. Thus, the hypothesis that parallelism results in media use behaviour is borne out in Poland. On the other hand, the distinct profile of social media users is confirmed in all three countries, with the partial exception of the Czech Republic, where social media load on the same factor as radio and TV.

7.5 Indicators of Alternative Narratives: Attitudes to Covid-19 and to the War in Ukraine

Previous research shows that there is a link between the Covid-19 narratives and the war in Ukraine (Padalko & Morita, 2023). It consists in some actors switching from expressing a sceptical view about the existence of C-19 and the effectiveness of vaccines to a narrative supporting Russia in the war in Ukraine. Evidence suggests that this process occurred on the supply side. In our research we hypothesized that a demand-side phenomenon may also have occurred, i.e. that people suspicious of the official narrative about C-19 were more susceptible to the Russian disinformation regarding the war in Ukraine.

In order to verify this hypothesis we constructed two composite indicators of attitudes. These are sets of statements on a Likert scale. The sceptical attitude towards the Covid-19 vaccination is operationalized as agreement with statements 2, 6, 7, 8, 9, 11, 12 and disagreement with statements 1, 3, 4, 5, 10, with the "don't know" answers in the middle of the scale. In order to normalize the distributions, a set of 12 dummy variables was created and a summary index operationalizing the attitude was calculated (Table 7.6).

Table 7.6 Opinions on Covid-19 vaccinations

Do you agree with the following statements?	*yes*	*no*	*don't know*
1. I support vaccination against Covid-19			
Poland	55%	31%	14%
Czech Republic	51%	43%	6%
Slovakia	45%	43%	12%
2. The Polish/Czech/Slovak government is using Covid-19 to strengthen its power			
Poland	50%	29%	21%
Czech Republic	41%	39%	20%
Slovakia	49%	34%	17%
3. Overall, I support vaccinating children against Covid-19			
Poland	44%	40%	15%
Czech Republic	30%	54%	15%
Slovakia	29%	56%	15%
4. I would agree to vaccinate my children against Covid-19			
Poland	40%	41%	18%

(*continued*)

Table 7.6 (continued)

Do you agree with the following statements?	*yes*	*no*	*don't know*
Czech Republic	23%	51%	26%
Slovakia	26%	53%	21%
5. Everyone should be vaccinated against Covid-19, no matter what type of vaccine			
Poland	37%	47%	17%
Czech Republic	28%	60%	11%
Slovakia	35%	48%	17%
6. Covid-19 vaccination carries more risks than benefits			
Poland	30%	48%	21%
Czech Republic	34%	48%	18%
Slovakia	41%	42%	17%
7. It is perfectly sufficient to get vaccinated with one or two doses of the Covid-19 vaccine			
Poland	25%	39%	36%
Czech Republic	34%	34%	32%
Slovakia	32%	39%	29%
8. Covid-19 is China's political weapon			
Poland	24%	40%	36%
Czech Republic	18%	52%	36%
Slovakia	19%	54%	27%
9. Vaccination against Covid-19 causes pregnancy problems			
Poland	17%	36%	47%
Czech Republic	31%	24%	45%
Slovakia	32%	31%	37%
10. If there was no other option, I would be vaccinated with the Sputnik vaccine			
Poland	12%	68%	20%
Czech Republic	13%	70%	18%
Slovakia	32%	49%	19%
11. Covid-19 testing is dangerous			
Poland	12%	68%	20%
Czech Republic	18%	67%	15%
Slovakia	29%	54%	17%
12. Covid-19 is a US political weapon			
Poland	11%	60%	29%
Czech Republic	17%	56%	27%
Slovakia	33%	44%	23%

"Strongly agree" and "moderately agree" combined; "strongly disagree" and "moderately disagree" combined

Table 7.7 Mean levels of Covid-scepticism in Poland, Czech Republic and Slovakia

Country	*Mean*	*N*	*SD*
Poland	6.32	1021	3.27
Czech Republic	6.47	1002	3.15
Slovakia	6.48	1013	3.20

Tukey's post-hoc test shows that all three pairs of differences are not significant

Attitudes towards the Covid-19 vaccines are similar in the three countries under study. The Czech and Slovak score is almost exactly the same, while the difference between the mean in Poland and the other two countries is not significant (Table 7.7).

Disinformation related to the conflict in Ukraine is operationalized as a set of statements consistent with our definition of disinformation, i.e. information distorting the reality with evil intentions. Items 1 to 8 reflect the narrative promoted by state-affiliated Russian actors. In order to control for other types of disinformation, we introduced items 9 and 10, operationalizing anti-Russian disinformation; however, we did not include them in the index construction, as they did not form the same attitude after initial scaling analysis. We selected items 1–8 and created a set of 8 dummy variables to normalize distributions. We then counted the summary index, operationalizing the attitude which can be described as susceptibility to Russian disinformation on the war in Ukraine (Table 7.8).

The pro-Russian attitudes related to the war in Ukraine are strongest in Slovakia and weakest in Poland, while Czech society occupies and intermediary position (Table 7.9).

The attitudes to Covid-19 and to the war in Ukraine are strongly correlated in all three countries (Table 7.10).

7.6 Media Use and Attitudes to Covid-19 and to the War in Ukraine

In the first step, bivariate analysis was conducted to establish how using particular news outlets is related to the strength of both attitudes. We obtained country-level hierarchies for media in regard to the strength the attitude. We divided the media outlets into categories in the presentation of results.

Table 7.8 Disinformation regarding the war in Ukraine

Do you agree with the following statements?	*yes*	*no*
1. The Ukrainian authorities are fascist or Nazi in character		
Poland	10%	90%
Czech Republic	26%	74%
Slovakia	37%	63%
2. The territory of today's Ukraine consists of historically Russian lands, and therefore Russia has the right to intervene there militarily		
Poland	12%	88%
Czech Republic	17%	83%
Slovakia	35%	65%
3. Ukraine's borders are not internationally recognized by the United Nations		
Poland	26%	74%
Czech Republic	37%	63%
Slovakia	38%	62%
4. There are US-funded laboratories in Ukraine that manufacture biological weapons		
Poland	18%	82%
Czech Republic	27%	73%
Slovakia	43%	57%
5. Russia is only defending its legitimate interests and bears no responsibility for the conflict in Ukraine		
Poland	16%	84%
Czech Republic	19%	81%
Slovakia	34%	66%
6. Russia's actions in Ukraine are a response to the crimes committed by Ukraine against the Russian-speaking population in the eastern part of the country		
Poland	17%	83%
Czech Republic	31%	69%
Slovakia	45%	55%
7. There are many criminals among the refugees from Ukraine		
Poland	38%	62%
Czech Republic	43%	57%
Slovakia	53%	47%
8. Refugees from Ukraine are treated in Poland/Czech Republic/Slovakia better than citizens of European Union Member States (e.g. they have easier access to public health care)		
Poland	72%	28%
Czech Republic	73%	27%
Slovakia	73%	27%

(*continued*)

Table 7.8 (continued)

Do you agree with the following statements?		
	yes	*no*
9. The Ukrainian armed forces have a military advantage over the Russian army		
Poland	26%	74%
Czech Republic	17%	83%
Slovakia	17%	83%
10. As part of the economic sanctions imposed on the Russian Federation, European Union Member States and the United States have suspended trade with the Russian Federation		
Poland	72%	28%
Czech Republic	66%	34%
Slovakia	56%	44%

"Strongly agree" and "moderately agree" merged; "strongly disagree" and "moderately disagree" merged

Table 7.9 Mean levels of disinformation related to the war in Ukraine in Poland, the Czech Republic and Slovakia

Country	*Mean*	*N*	*SD*
Poland	3.27	1021	2.10
Czech Republic	3.99	1002	2.33
Slovakia	4.71	1013	2.49

Tukey's post-hoc test shows that all three pairs of differences are significant at $p<0.001$

Table 7.10 Correlation between Covid-scepticism and disinformation related to the war in Ukraine indicators

Country	*Pearson's R*	*Sig.*
Poland	0.470	0.000
Czech Republic	0.430	0.000
Slovakia	0.583	0.000

In Poland, as far as print media are concerned, there is significantly higher pro-Russian sympathy among the readers of some (but not all) right-wing titles. However, this does not apply to Covid-scepticism. The only newspaper with readers that are both more pro-Russian and Covid-sceptical than average is the tabloid Fakt. At the other end of the scale, readers of newspapers and magazines such as Polityka, Newsweek and

Dziennik Gazeta Prawna are marked by below-average strength of both attitudes.

In the case of radio and TV stations, it is important to note that none have audiences that would show above-average strength of either attitude to a significant degree. On the other hand, audiences of outlets such as TVN, Polsat, TV Trwam and TOK FM show significantly lower strength of attitudes to C-19 and Ukraine. These stations are from the opposite sides of the political spectrum in terms of their editorial policy, which suggests that C-19 and the war in Ukraine do not polarize in Poland to a significant degree. On the other hand, users of social media platforms display above-average values of the pro-Russian attitude (Snapchat, Instagram), or both (Facebook, YouTube, TikTok). Only people using LinkedIn for news are on the opposite side of the scale among social media users.

In the Czech Republic, print media readers are either at or below the average level, both in the case of attitudes to C-19 and to the war in Ukraine (the above-average values are not statistically significant). On the other hand, readers of MF Dnes, Hospodářské noviny, Respekt, E15 and Forbes show below-level values for both indicators. The same can be noted for internet websites associated with print media.

As regards the social media platforms, users of TikTok, Facebook and Pinterest are far more sceptical about C-19 vaccines than the average Czech. This regularity, however, does not apply to attitudes regarding the war in Ukraine. Social media users share a distinct socio-demographic profile (i.e. they are much younger than average), which may explain the observed regularity. The multivariate analysis indeed indicates the role of age in predicting attitudes to C-19.

In Slovakia, the readers of Báječná žena and Slovenka are especially prone to pro-Russian attitudes, while those who read Hospodárske noviny, Sme and Denník N are at the opposite end of the spectrum. It is interesting to note that no such differences exist with regard to C-19 scepticism. Listeners of Rádio Patria and Rádio Regina are particularly susceptible to positive reception of Russian narratives related to the war in Ukraine. On the other hand, the audiences of two state TV channels are at the opposite end of the continuum. People who find their information on TikTok and Facebook are particularly prone to be receptive to both counter-narratives under study (Table 7.11).

Various media outlets are used in different combinations, not necessarily the same in the three countries under study. Moreover, other factors

Table 7.11 Attitudes and the use of media to obtain information

PL	N	UA (+/-)	sig.	Covid (+/-)	sig.	CZ	N	UA (+/-)	sig.	Covid (+/-)	sig.	SK	N	UA (+/-)	Sig.	Covid (+/-)	Sig.
Print																	
Gazeta Polska	72	1.03	*	0.41		Blesk	74	0.30		0.21		Báječná žena	68	0.59	*	0.56	
Nasz Dziennik	47	0.8	*	0.62		5plus2	49	0.26		-0.47		Slovenka	56	0.56	*	0.33	
Fakt	221	0.71	*	0.44	*	Právo	82	0.10		-0.37		Život	80	0.27		-0.07	
Super Express	186	0.63	*	0.51		Téma	40	-0.13		-0.26		Nový čas pre ženy	73	0.22		0.03	
G. Wyborcza	266	-0.01		-0.62	*	Lidové noviny	97	-0.34		-0.52		Plus 1 deň	79	0.12		-0.06	
Rzeczpospolita	116	-0.01		-0.6	*	Metro	83	-0.38		-0.66	*	Új Szó	28	0.02		0.19	
wSieci	53	-0.22		-0.38		MF Dnes	138	-0.39	*	-0.63	*	Korzár	61	0.01		-0.19	
Local / regional	148	-0.36	*	-0.62	*	Deník	124	-0.39	*	-0.30		Plus 7 dní	118	0.00		-0.61	*
Newsweek	137	-0.45	*	-1.54	*	Sport	76	-0.60	*	0.32		Zdravie	101	-0.13		-0.34	
Dziennik GP	109	-0.48	*	-0.88	*	Ekonom	38	-0.69	*	-0.79		Nový čas	124	-0.14		-0.40	
Tyg. Polityka	85	-0.77	*	-1.67	*	Hospodářské noviny	83	-0.72	*	-1.15	*	Pravda	129	-0.32		-0.71	*
Do Rzeczy	42	-0.79	*	0.69		Respekt	62	-0.79	*	-1.01	*	Hospodárske noviny	112	-0.69	*	-1.13	*
						Reflex	80	-1.05	*	-0.58		Sme	143	-0.93	*	-1.01	*
						E15	51	-1.11	*	-1.43	*	Denník N	109	-1.09	*	-1.14	*
						Forbes	33	-1.67	*	-1.55	*						
Radio and TV																	
Radio Maryja	39	0.29		-0.78	*	Rádio Čas	38	0.72	*	0.82		Rádio Patria	51	0.60	*	0.91	*
TVP (public)	308	0.01		0.07		Barrandov TS	107	0.52	*	-0.23		Rádio Regina	116	0.40	*	0.27	
RMF FM	548	-0.04		-0.03		Country rádio	62	0.51	*	-0.78		Rádio Jemné	160	0.13		0.44	*
Radio Zet	407	-0.17	*	-0.18		Fajn Radio	65	0.49	*	0.91	*	Rádio Vlna	181	0.09		0.13	
Polsat	514	-0.2	*	-0.26	*	Rádio Kiss	113	0.24		0.68	*	Rádio Európa 2	176	0.03		0.35	
PR (public)	150	-0.24		-0.45	*	Rock Rádio	72	0.04		-0.11		TV Lux	72	-0.02		0.17	
TVN	609	-0.27	*	-0.48	*	Rádio Blaník	124	0.00		-0.05		Rádio Slovensko	305	-0.19	*	-0.83	*
TV Republika	50	-0.52	*	-0.23		FTV Prima	452	-0.09		-0.24	*	TV JOJ	402	-0.21	*	-0.27	*
Telewizja Trwam	39	-0.57	*	-0.84	*	Rádio Impuls	205	-0.10		-0.19		TA3	386	-0.23	*	-0.60	*
Radio TOK FM	110	-0.59	*	-1.36	*	Nova Group	425	-0.16	*	-0.30	*	Fun rádio	202	-0.23		0.03	
						Hitrádio	96	-0.21		0.14		Rádio Expres	340	-0.29	*	-0.45	*
						Frekvence 1	179	-0.27	*	-0.29		TV Markíza	405	-0.30	*	-0.36	*
						Seznam.cz TV	299	-0.28	*	-0.40	*	Dvojka	229	-0.34	*	-0.67	*
						Evropa 2	170	-0.32	*	0.30		Jednotka	423	-0.38	*	-0.76	*
						Česká televize	588	-0.51	*	-0.64	*						
						Rádio Beat	67	-0.66	*	-1.02	*						
						Český Rozhlas	269	-0.99	*	-1.34	*						
Internet portals																	
Niezależna.pl	80	-0.03		0.4		Blesk.cz	92	0.25		-0.16		dnes24.sk	151	0.29	*	0.16	

Independent-samples T-test. Reference group: non-users of the medium. Shaded cells: difference of means significant at $p<0.05$ or higher

The values are differences in the average level of disinformation of consumers of a given medium from the national mean

A positive sign indicates above-average disinformation, while a negative sign indicates a below-average level such as demographics (gender and age), socio-economic position (education and material living conditions) and place of residence may have an influence irrespective of the media use, but through structural factors. For instance, older people may be less susceptible to negative attitudes to Covid-19 vaccines due to being more vulnerable. Therefore, multiple regression models were created in order to control for these factors. All models were constructed separately for Poland, the Czech Republic and Slovakia.

In the first model, we determined the net influence of socio-demographic factors on attitudes to Covid-19 vaccines. In Model 2, we introduced the variables related to media consumption patterns. Please note that the components in the factor scores differ between countries, while the socio-demographic variables are coded in the same way.

In Models 3, 4, and 5, the dependent variable is susceptibility to Russian disinformation related to the war in Ukraine. Model 3 contains the socio-demographics and Model 4 introduces the media consumption variables.

Model 5 is the summary of our analysis. Our theory assumes that the attitudes towards C-19 (the first stage of the crisis) were carried over to the attitudes related to the war in Ukraine, a view strengthened by the observation of some actors spreading disinformation. There was a path. Therefore, in the final model, the attitude to the earlier crisis (Covid-scepticism) was included as an explanatory variable to verify its significance.

C-19 scepticism is strongly predicted by socio-demographic factors in all three countries under study (Table 7.12). Women are more prone to this attitude than men. Higher levels of distrust to vaccines are typical of less educated people, residing in villages or small towns, in poor material conditions. Negative attitudes to Covid vaccinations decrease with age. Thus, a typical Covid-sceptic would be a poor, young and uneducated woman living in a small town or a village.

When media use is introduced into the equation, differences between the three countries become apparent. In Poland, Factor 1 (use of state/pro-government media) reduces the levels of vaccine scepticism. In the Czech Republic, Factor 2 is significant, i.e. respondents getting their news from main internet portals are less likely to be sceptical. In Slovakia, using the main portals or the electronic media reduces Covid-scepticism, while the use of social media is associated with higher levels of this attitude.

The socio-demographic predictors of attitudes to the war in Ukraine are different from the factors influencing attitudes to C-19, and differ

Table 7.12 Multiple regression analysis. Dependent variable: Covid-scepticism

Dependent variable: Covid-scepticism	Beta	Sig.	Beta	Sig.	Beta	Sig.
Model 1						
	PL		CZ		SK	
Gender (1=woman)	0.173	0.000	0.156	0.000	0.113	0.000
Education (1=post-secondary)	-0.098	0.001	-0.123	0.000	-0.177	0.000
Residence (1=100K or more)	-0.081	0.006	-0.175	0.000	-0.078	0.010
Material living conditions (1=good)	-0.122	0.000	-0.075	0.016	-0.161	0.000
Age	-0.263	0.000	-0.157	0.000	-0.039	0.200
R-sq	0.154		0.098		0.087	
Model 2						
	Beta	Sig.	Beta	Sig.	Beta	Sig.
	PL		CZ		SK	
Gender (1=woman)	0.166	0.000	0.139	0.000	0.119	0.000
Education (1=post-secondary)	-0.086	0.004	-0.093	0.003	-0.158	0.000
Residence (1=100K or more)	-0.083	0.005	-0.147	0.000	-0.080	0.008
Material living conditions (1=good)	-0.110	0.000	-0.060	0.051	-0.159	0.000
Age	-0.228	0.000	-0.178	0.000	-0.005	0.866
Factor score 1*	-0.134	0.000	-0.017	0.574	-0.135	0.000
Factor score 2	0.010	0.716	-0.186	0.000	-0.051	0.087
Factor score 3	0.033	0.324	0.008	0.802	0.116	0.000
R-sq	0.171		0.129		0.120	

[a]Components of factor scores differ between countries. They are listed for each country in Tables 7.3, 7.4 and 7.5

between countries. In Poland, gender, age and education are significant: young women with low education show above-average levels of this attitude. In the Czech Republic, on the other hand, place of residence in addition to age is significant. In Slovakia, higher values are predicted by gender (women are more susceptible), education, place of residence, material conditions and age. It is important to note the different direction of the influence of age on attitudes: in Poland, older people are less likely to have pro-Russian sympathies, while in the Czech Republic and Slovakia, they are more likely to.

Media use influences vulnerability to Russian disinformation differently to the way it affects vaccine scepticism. In Poland, the use of the state and pro-government media reduces susceptibility to Russian disinformation, while social media increases it. In the Czech Republic, using the internet decreases susceptibility. In Slovakia, as in the Czech Republic, people who gather their information from internet portals are less susceptible. Like in Poland, Slovaks who use social media are more likely to be affected by this type of information distortion.

Finally, in Model 5, we test the hypothesis about path dependency, i.e. the carry-over effect of C-19 pandemic attitudes to attitudes regarding the war in Ukraine. In all three countries, vaccine-scepticism is the strongest predictor of vulnerability to the Russian narrative, as shown by the high values of the Beta coefficients for the Covid variable in Model 5 for all countries (Table 7.13).

Table 7.13 Multiple regression analysis. Dependent variable: susceptibility to Russian disinformation related to the war in Ukraine

Dependent variable: susceptibility to Russian disinformation related to the war in Ukraine	Beta	Sig.	Beta	Sig.	Beta	Sig.
Model 3						
	PL		CZ		SK	
Gender (1=woman)	0.118	0.000	0.028	0.358	0.091	0.002
Education (1=post-secondary)	-0.177	0.000	-0.170	0.000	-0.160	0.000
Residence (1=100K or more)	-0.015	0.626	-0.109	0.000	-0.124	0.000
Material living conditions (1=good)	-0.053	0.075	-0.104	0.001	-0.147	0.000
Age	-0.231	0.000	0.069	0.029	0.109	0.000
R-sq	0.127		0.077		0.100	
Model 4						
	Beta	Sig.	Beta	Sig.	Beta	Sig.
	PL		CZ		SK	
Gender (1=woman)	0.107	0.000	0.006	0.835	0.096	0.001
Education (1=post-secondary)	-0.168	0.000	-0.133	0.000	-0.144	0.000
Residence (1=100K or more)	-0.014	0.630	-0.074	0.016	-0.125	0.000
Material living conditions (1=good)	-0.053	0.080	-0.084	0.006	-0.145	0.000
Age	-0.168	0.000	0.045	0.151	0.141	0.000
Factor score 1*	-0.073	0.014	-0.003	0.932	-0.105	0.000
Factor score 2	0.016	0.576	-0.236	0.000	-0.011	0.714
Factor score 3	0.112	0.001	0.028	0.362	0.120	0.000
R-sq	0.141		0.128		0.124	
Model 5						
	PL		CZ		SK	
Gender (1=woman)	0.042	0.131	-0.049	0.081	0.032	0.206
Education (1=post-secondary)	-0.134	0.000	-0.096	0.001	-0.059	0.022
Residence (1=100K or more)	0.018	0.507	-0.016	0.579	-0.081	0.001
Material living conditions (1=good)	-0.009	0.738	-0.061	0.033	-0.059	0.021
Age	-0.078	0.020	0.115	0.000	0.144	0.000
Factor score 1*	-0.020	0.463	0.004	0.883	-0.033	0.193
Factor score 2	0.012	0.651	-0.163	0.000	0.017	0.508
Factor score 3	0.099	0.002	0.025	0.377	0.057	0.028
Covid	**0.394**	**0.000**	**0.394**	**0.000**	**0.539**	**0.000**
R-sq	0.270		0.264		0.380	

[a]Components of factor scores differ between countries. They are listed for each country in Tables 7.3, 7.4 and 7.5

7.7 Verification of the Hypotheses and Discussion

Hypothesis 1 concerning the patterns of media use is confirmed in Poland. The regularity consists in using predominantly either state media, major private outlets or social media to obtain information. Moreover, state media use was associated with tuning in to pro-government private outlets. This pattern was characteristic of people who supported the PiS government at the time. On the other hand, in both the Czech Republic and in Slovakia, reading news in print constituted a separate factor. The distinctness of social media users is confirmed in all three countries.

Hypothesis 2 concerned the latter group, i.e. social media users, as a category particularly vulnerable to alternative narratives. In the case of Covid-19, this hypothesis is confirmed in Slovakia. As regards the pro-Russian attitude, it is true in Poland and Slovakia. This means that in Slovakia, social media users are particularly vulnerable to disinformation activities.

Hypothesis 3 concerned between-country differences as regards the two attitudes. We assumed that they would be pronounced in the case of the war in Ukraine. The C-19 attitudes are very similar in all three countries. However, attitudes to the war in Ukraine differ between countries. In Poland, pro-Russian narratives met with the weakest response, while Slovaks are the most susceptible to this type of influence.

Hypothesis 4 is confirmed in all three countries. The carry-over effect is strong and is a significant contribution to the explanatory power of the models in all countries. Developing scepticism to the elite consensus during the first crisis persisted during the war in Ukraine. People who had been sceptical about C-19 policies were also more likely to agree with the pro-Russian narratives regarding the war.

Our initial hypotheses about the origin of the spread of alternative narratives have been confirmed to varying degrees. On the one hand, the carry-over hypotheses is documented: the C-19 formed a basis for a durable attitude that manifested itself in the rejection of the mainstream, consensus interpretation of events. On the other hand, the relation between media consumption and attitudes to C-19 and to the war in Ukraine is relatively weak. We expected both attitudes to be strongly linked to using social media for information. However, this relation is not universal. The effect applies to Slovakia more than to the other countries. It is by no means solely social media users who reject the official narratives and subscribe to alternative views. Therefore, while allowing for some influence of deliberate disinformation activities through these platforms, we must look for other possible explanations. Here are some of them, presented as ideas for further study.

Consensus breeds dissent. A state-supported narrative involves give-and-take. Society is protected from harmful content and disinformation, but this process involves curtailing rights such as free speech (Jain, 2022; Kalitenko et al., 2021) and freedom of movement (Simoes, 2021). In this sense, it violates the principles of democracy. This is particularly important if the restrictions are performed without sufficient evidence, or the evidence is ex-post invalidated. Therefore, we may hypothesize that the rejection of consensus on C-19 may have stemmed from distrust of the state caused by what some people saw as unwarranted intervention into personal freedom. As such, it is not necessarily a manifestation of irrational beliefs, but on the contrary, of rational calculation of risks. For instance, we noted that with age, Covid-scepticism diminishes. Perhaps the reason for such a finding is the lower vulnerability of younger people to the risks posed by the disease. They had less to gain and more to lose by the C-19 related policies, especially the lockdowns.

In the case of the war in Ukraine, views consistent with the Russian narrative may be more common in Slovakia and the Czech Republic than in Poland due to differences in the overall geopolitical orientation of the population. This is reflected in attitudes to NATO, to the EU and to the international role of the state. Poles are among the most supportive of their country's membership in NATO, while Slovaks are among the least in favour (NATO, 2023). Attitudes to Russia were especially positive in Slovakia even before the war started (Pew Research Center, 2019). This orientation of Slovak society was reflected in, and reinforced by, government policy. Further research should help explain to what extent deliberate actions of the Russia-related actors helped contribute to an attitude change. Our data suggests that the views of populations shaped by historical legacy and collective memory are the dominant predictors of attitudes to the war.

References

Allport, G. W. (1929). The composition of political attitudes. *American Journal of Sociology, 35*(2), 220–238.

Bajomi-Lázár, P. (2014). *Party colonisation of the media in Central and Eastern Europe.* Central European University Press.

Crilley, R., & Gillespie, M. (2019). What to do about social media? Politics, populism and journalism. *Journalism, 20*(1), 173–176.

De Albuquerque, A. (2018). Political parallelism. In *Oxford research encyclopedia of communication.*

Dębiec, K. (2021). *The Bohemianisation of the media. The acquisition of the Czech media sector by local billionaires.* Ośrodek Studiów Wschodnich. Retrieved June 10, 2024, from https://www.osw.waw.pl/en/publikacje/osw-commentary/2021-02-11/bohemianisation-media-acquisition-czech-media-sector-local

Dobek-Ostrowska, B. (2019). How the media systems work in Central and Eastern Europe. In E. Połońska & C. Beckett (Eds.), *Public service broadcasting and media systems in troubled European democracies* (pp. 259–278). Springer.

Dobek-Ostrowska, B., & Glowacki, M. (2015). *Democracy and media in Central and Eastern Europe.* Peter Lang.

Doroshenko, L., & Lukito, J. (2021). Trollfare: Russia's disinformation campaign during military conflict in Ukraine. *International Journal of Communication, 15*, 28.

Dragomir, M. (2020). Media influence matrix: Slovakia. In *Funding journalism.* CMDS.

Freedom House. (2024) Poland: Country profile. Retrieved June 10, 2024, from https://freedomhouse.org/country/poland

Geers, S. (2020). News consumption across media platforms and content: A typology of young news users. *Public Opinion Quarterly, 84*(S1), 332–354.

Hallin, D. C., & Mancini, P. (2004). *Comparing media systems: Three models of media and politics.* Cambridge University Press.

Hoyle, A., Powell, T., Cadet, B., & van de Kuijt, J. (2022). Web of lies: Mapping the narratives, effects, and amplifiers of Russian Covid-19 disinformation. In *COVID-19 disinformation: A multi-national, whole of society perspective* (pp. 113–141). Springer International Publishing.

Jain, A. (2022). Restricting freedom of speech: An analysis of censorship cases in relation to misinformation during the COVID-19 crisis. *Brandeis University Law Journal, 10*(1).

Jakubowicz, K. (2001). Rude awakening social and media change in Central and Eastern Europe. *Javnost-The Public, 8*(4), 59–80.

Kalitenko, O., Anikinar, G., Spasova, E., & Shahaka, O. (2021). The restrictions of the freedom of information during the Covid-19 pandemic. *Cuestiones Políticas, 39*(70).

Khaldarova, I., & Pantti, M. (2020). Fake news: The narrative battle over the Ukrainian conflict. In *The Future of Journalism: Risks, Threats and Opportunities* (pp. 228–238). Routledge.

Kovacheva, R. (2023). Disinformation on COVID-19 and the war in Ukraine: shared narratives and channels of. In *КОМУНИКАЦИЯТА И МЕДИИТЕ НА XXI ВЕК: ОБРАЗОВАТЕЛНИ И ПРОФЕСИОНАЛНИ ПРЕДИЗВИКАТЕЛСТВА: Международна научна конференция на ФЖМК*, проведена на 27–28.10. 2022 г. (pp. 34–51). Факултет по журналистика и масова комуникация, Софийски университет "Св. Кл. Охридски".

Lawrence, M., Homer-Dixon, T., Janzwood, S., Rockstöm, J., Renn, O., & Donges, J. F. (2024). Global Polycrisis: The causal mechanisms of crisis entanglement. *Global Sustainability, 7*, e6.

Milewski, D. (2020). The analysis of narratives and disinformation in the global information environment amid Covid-19 pandemic.

NATO. (2023) NATO Audience Research: Pre-Summit polling results 2023. Retrieved June 13, 2024, from https://www.nato.int/nato_static_fl2014/assets/pdf/2023/7/pdf/2300707-pre-summit-research-2023.pdf

Nicoli, F., van der Duin, D., Beetsma, R., Bremer, B., Burgoon, B., Kuhn, T., & de Ruijter, A. (2024). Closer during crises? European identity during the COVID-19 pandemic and the Russian invasion of Ukraine. *Journal of European Public Policy*, 1–27.

Padalko, H., & Morita, P. (2023). Covid-19 'infodemic' management: Russia's key tool for influencing geopolitics. *Population Medicine, 5*, A1809. https://doi.org/10.18332/popmed/164626

Pew Research Center. (2019). Global Attitudes Survey. Retrieved June 13, 2024, from https://www.pewresearch.org/short-reads/2020/02/07/russia-and-putin-receive-low-ratings-globally/

Połońska, E. (2019). Watchdog, lapdog, or attack dog? Public service media and the Law and Justice government in Poland. In E. Połońska & C. Beckett (Eds.), *Public service broadcasting and media systems in troubled European democracies* (pp. 227–255). Springer.

Rzeczkowski, G., Witkowski, P., & Zarzycki, R. (2024). Analiza narracji prawicowo-populistycznych ruchów antyestablishmentowych w mediach społecznościowych w odniesieniu do szczepień i wojny w Ukrainie. In K. Stasiuk-Krajewska & M. Wenzel (Eds.), *Dezinformacja w czasach kryzysu* (pp. 141–164). Adam Marszałek.

Shu, K., Sliva, A., Wang, S., Tang, J., & Liu, H. (2017). Fake news detection on social media: A data mining perspective. *ACM SIGKDD Explorations Newsletter, 19*(1), 22–36.

Simoes, F. D. (2021). COVID-19 and International Freedom of Movement: A Stranded Human Right? *Yale Journal of Health Policy, Law, and Ethics, 20*, 362.

Spadaro, A. (2020). COVID-19: Testing the limits of human rights. *European Journal of Risk Regulation, 11*(2), 317–325.

Stetka, V. (2012). From multinationals to business tycoons: Media ownership and journalistic autonomy in Central and Eastern Europe. *The International Journal of Press/Politics, 17*(4), 433–456.

Stjernfelt, F., & Lauritzen, A. M. (2020). Facebook and Google as offices of censorship. In *Your post has been removed: Tech giants and freedom of speech* (pp. 139–172). Springer.

Terracciano, B. (2023). Accessing to a "Truer Truth": Conspiracy and figurative reasoning from Covid-19 to the Russia-Ukraine war. *Media and Communication, 11*(2), 64–75.

Wenzel, M. (2018). News consumption and political party preferences in Poland. *Political Preferences, 18*, 57–74.

Wiik, J. (2019). *Professionalism, professionalization, and deprofessionalization in journalism*. Oxford Research Encyclopedia of Communication.

CHAPTER 8

Comparing Electoral Interference Perceptions in Czechia, Poland, and Slovakia

Nik Hynek, Vaclav Moravec, Beata Gavurova, and Viliam Kovac

8.1 Introduction

The nature of late—or, better still, liquid—modern liberal-democratic societies has undergone significant change in socio-political and technological respects. In the socio-political sphere, there is an ongoing debate

This paper has received funding from the European Union CEF-TC-2020-2, contract number: 2020-EU-IA-0267—Central European Digital Media Observatory (CEDMO).

N. Hynek
Department of Security Studies, Faculty of Social Sciences, Charles University in Prague, Prague, Czechia
e-mail: hynek@fsv.cuni.cz

V. Moravec
Department of Journalism, Faculty of Social Sciences, Charles University in Prague, Prague, Czechia
e-mail: vaclav.moravec@fsv.cuni.cz

M. Wenzel, K. Stasiuk-Krajewska (eds.), *Disinformation in Central Europe*, https://doi.org/10.1007/978-3-031-76871-2_8

on the crisis of democracy, particularly its critical infrastructure (Müller, 2022), its representativeness (Urbinati, 2016), and its legitimacy (Dahrendorf, 2003), as well as its preparedness for crises in relation to social cohesion (Anheier and Filip, 2021; Pérez-Curiel & Rivas-de-Roca, 2023; 2021). We can observe the emergence of political marketing and the expanding cartelization of party systems (Katz & Mair, 1995, Katz & Mair, 2009), alongside the mainstreaming of domestic populism (Kriesi, 2018; Schwörer, 2021; Vachudova, 2021), which extends into the international system as a new form of political-security revisionism.

Liquid modernity (Bauman, 2000) represents a general framework for understanding the deep transformations of politics and contemporary society alike, emphasizing the fluid and unpredictable characteristics of social, economic, and cultural processes. The state of crisis, to paraphrase the title of Bauman and Bordoni's (2016) influential book, has become the new normal. Liquid modernity thus creates a space for investigating the challenges and opportunities in a dynamically altering world. The current period can be characterized by insatiable consumption, mobility, contingent social and political relations, and strategic flexibility and hedging. Digitalization brings new meanings to modernity, making human relationships more fluid with faster, easier connectivity (Lee, 2022). Databases have become new power locations in digital modernity, influencing knowledge production, management, and capital (O'Hara, 2020). Bauman and Haugaard (2008) noted, over a decade ago, that liquid modernity's characteristics have been changing, with social class also evolving as a key sociological tool (Atkinson, 2008).

Bauman's liquid modernity has been a significant influence among political sociologists (Ercan & Hendricks, 2013; Beck, 2018; Elliott & Lemert, 2021), consumer culture theorists (Gabriel & Lang, 2018; Lury, 2020), as well as political economists (Mason, 2016; Standing, 2021). All emphasize that the breakdown of old social structures allows for the

B. Gavurova (✉)
Center for Applied Economic Research, Faculty of Management and Economics, Tomas Bata University in Zlin, Zlín, Czechia
e-mail: gavurova@utb.cz

V. Kovac
Technical University of Košice, Košice, Slovakia
e-mail: viliam.kovac@tuke.sk

emergence of unprecedented types of crises, more diverse lifestyles as well as heightened insecurities (Caldwell & Henry, 2020; Lagna & Lenglet, 2020). Liquid journalism respects the rights and privileges of individuals as both creators and users of their news (Deuze, 2008), redefining journalism and investigative reporting (Wall, 2017; Cancela & Dubied, 2022). Bauman highlights the impact of major powers on late modern society and the decline in centralized nation-state power, creating chaos in modernity (Sabri, 2021). Bauman's concept is also applied in other areas, such as liquid neutrality in security strategies (Roitto & Holmila, 2021) and liquid democracy in election campaigns (Vestri, 2015).

The rise of domestic political populism (Albertazzi & McDonnell, 2008; Mudde & Kaltwasser, 2017; Norris & Inglehart, 2019) as well as foreign and security policy revisionism (Allison, 2017; Kagan, 2018; Götz & Merlen, 2019) further complicates the landscape of modern democracies. Populist movements often exploit the fluidity and uncertainty of liquid modernity, using it to challenge established political norms and institutions. These movements capitalize on widespread discontent and anxiety, offering simplistic solutions to complex problems, thereby gaining significant traction. Populism thrives in an environment where traditional political affiliations and ideologies are increasingly questioned, and where the electorate is more susceptible to emotional and sensational appeals. This phenomenon is particularly relevant in the context of our study, as it examines how perceptions of foreign interference and political party support are shaped in such a volatile environment. Foreign and security policy revisionism, on the other hand, refers to the efforts by major powers to challenge the existing international order, and to assert their influence in ways that destabilize regional and global security. This includes tactics such as electoral interference, propaganda, and cyber warfare, which are designed to undermine democratic processes and to sow discord within and between nations. Understanding public perception of the interplay between domestic political populism and foreign security revisionism is crucial for developing effective foreign and security policies that can withstand these challenges.

A specific feature of central Europe, that is, the region we empirically focus on, is its frequent reaction to previous institutional-mimetic processes of euro-atlantic integration and westernization (Holmes and Krastev, 2018; Zielonka & Rupnik, 2020). The outcomes of these phenomena include illiberal democracy, defective democracy, and processes of de-democratization (Bogaards, 2018; Mudde, 2021; Tilly, 2023), often supported by alternative epistemic configurations, whether neo-nationalist, Marxist, or directive-managerial (Beck, 1996; Douglass, 2021; Paternotte & Verloo, 2021; Shaw, 2022). Contrasting the populist manipulation of citizens' attitudes is the debate about the role of experts and epistemic networks in determining societal priorities and policy solutions (Radaelli & O'Connor, 2009). Ganuza and Font (2020) argue that technocratic governance is a popular alternative to representative democracy, especially in countries exhibiting a high dissatisfaction with democracy and a tradition of bureaucratic governments.

The perceived influence of major powers on domestic elections is an issue that remains underexplored in general, and within liquid modernity in particular, thus representing a significant research gap. In times of democratic crises, attention to countries' foreign and security policies, including the public perception of threats from major powers, is crucial. This study examines regional partnership importance and the relationships between political party support and perceived major power influence on electoral processes in Czechia, Poland, and Slovakia. No previous studies have analyzed these perceptions through demographic characteristics and political party preferences. This research confirms/demonstrates the fluidity, instability, fragmentation, risk, and uncertainty of liquid modernity, revealing demographic profiles essential for developing digital and media literacy systems.

This study reflects these concerns and creates a robust research platform for investigating the selected aspects of domestic politics and foreign and security policy. By analyzing the perceived relationships between political party preferences and the influence of major powers on electoral processes in Czechia, Poland, and Slovakia, this research provides valuable comparative insights into how different demographic groups perceive security threats related to major foreign powers. The analytical processes are based on regression analysis using the R software environment, supplemented by additional packages. The findings highlight the importance of examining these aspects to identify not only concerns about interference by major powers in democratic processes, but also to gauge the level of digital and media literacy, vulnerability, and manipulability of the population.

8.2 Literature Review

8.2.1 *State of the Art: Electoral Interference by Major Powers*

Foreign and security policies are currently explored in various horizontal and vertical trajectories. Foreign interference in elections has become an increasingly important tool for influencing politics in other countries. However, few studies examine how different population groups perceive interference by major powers in order to identify threats, risks, vulnerabilities, and population literacy. In the context of the defined main goal, this study examines the existing literature on electoral interferences by major powers, their methods and effects.

Technological development has radically affected all aspects of human life and society in recent years. New tools can measure human behavior and predict human decisions. Artificial intelligence, machine learning, intelligent algorithms, and autonomous robots have gained tremendous power to engage with voters in election processes by creating voter psychographic and behavioral profiles. Various advertising campaigns profile voters or groups in order to influence elections. Kamarck (2018) states that the Internet can mobilize large numbers of people for political activities, generate false information about candidates, influence voter lists and election processes, and reduce willingness to vote. Social media can become a tool for democracy and against it. For effective campaigns on social networks, the timing and visual content of posts, as well as interactivity, are crucial (Koc-Michalska et al., 2021).

According to Levin (2019), partisan election intervention occurs when one or more sovereign countries deliberately carry out activities to influence upcoming elections in another sovereign country, either openly or covertly. This act is done intentionally to help or harm one of the parties, and can result in significant costs—both immediate and long-term—and potentially cause permanent damage to relationships. Until 2011, there had been limited theoretical and empirical research on the influence of foreign electoral interventions (Sethi, 2007). Research outcomes indicate that between 1946 and 2000, the United States and the Soviet Union (and later the Russian Federation) intervened 117 times with the aim of influencing election results—81 times by the United States and 36 times by the Soviet Union/Russian Federation, with 80 of these interventions being covert. Agents of these electoral interventions include international institutions, non-governmental organizations, and individuals.

There are several ways of interfering in electoral processes that need detailed investigation to quantify the vulnerability level of political systems in a country. Au-Yeung and Tagirova (2024) mention the manipulation of party membership as one possibility. Political decisions create external economic and political costs. According to Aidt et al. (2021), foreign influence represents a strategic choice aimed at internalizing these externalities. This can manifest as voluntary agreements, political interventions based on rewarding and sanctioning the target country, and institutional interventions. Tomz and Weeks (2020) find that even minimal interventions polarize the public along party lines. Their research shows that the U.S. population is more likely to criticize foreign involvement, lose faith in democracy, and seek punishment if a major power supports the political opposition rather than helping their own political party. Political parties are key institutions in election processes and political systems. They lead and finance campaigns, organize public forums, propose policies, and mobilize their supporters. Non-governmental institutions, such as civil societies, support voter participation and monitor elections. Foreign actors can exploit digital communication technologies to hack databases of political parties and electoral bodies.

Many political scientists have examined the population's knowledge of policies and public issues (Giugni & Grasso, 2018; Clemens, 2021). Concerns about threats to democracy due to disinformation and disinformation campaigns have been investigated in relation to elections in various countries (Schia and Gjesvik, 2020; McKay and Tenove, 2021; Schünemann, 2022). News organizations can also play a significant role in election interference by importing political agendas into mainstream journalism. International news agencies can openly or covertly support or oppose candidates and parties. Online political advertising, whose effects can depend on foreign funding, is another crucial form of interference. This funding is difficult to control as most democracies lack legal regulations for it (Hiltunen, 2021; Bene et al., 2023). Governmental and non-governmental entities can use disinformation campaigns and social media accounts to influence and manipulate political sentiment to achieve geopolitical outcomes. Not only artificial intelligence, machine learning tools, and social media, but also political espionage actively develop disinformation processes (Pope, 2018; Deibert, 2022). Non-profit organizations also have the potential to influence election results within the framework of achieving social well-being goals. The scope and power of these tools are still insufficiently explored, and the risks are growing. Therefore, democracies must implement regulations to protect their electoral processes from interference by major powers. Nowadays, it is already possible to obtain information about persons and groups spreading harmful propaganda and disinformation.

Democratic systems and institutions are increasingly under attack from forces seeking to divide or destabilize them. Consequently, various European Union institutions are striving to develop tools to combat interference in national democracies, including platform regulation and industry oversight to address hybrid threats and disinformation (Schlipphak & Treib, 2017; Nicolaidis & Youngs, 2023).

Wigell (2019) recommends distinguishing conceptually between hybrid warfare and hybrid interference. Hybrid interference employs three main tools: covert diplomacy, geoeconomics, and disinformation. Western democracies are particularly vulnerable to hybrid interference, which exploits their liberal values to create societal divisions and weaken governance (Noor and Niaz, 2023). The Russian Federation's hybrid warfare against the North Atlantic Treaty Organization in the Baltic region exemplifies how soft power tools, such as disinformation and social media bots, can be effectively used to counter international military alliances (Shuker & Topor, 2021). It is also crucial to examine the ideological and geopolitical goals behind such interference. Gvosdev (2019) argues that when Russian activities are guided more by geopolitical interests than ideology, open interference in major non-Western democracies brings fewer benefits and poses significant risks of deteriorating mutual relations. Kondratov and Johansson-Nogués (2023) note that while Russia's hybrid interference may create or exacerbate societal turbulence in the short term, it can lead to a breakdown of vertical and horizontal trust in society in the long term.

Although the governments of the United States and Australia refer to interference as foreign, European institutions use different terminology (Berzina & Soula, 2020). According to Von der Leyen (2019), interference by other countries or major powers is termed "external interference." Academic sources have not yet reached a consensus definition of interference, leaving the distinction between interference and influence unclear. Therefore, defining narrower terms such as "unacceptable influence" and "illegitimate influence" in the information sphere could help delineate the framework of interference. This approach could extract critical elements to form a broader definition that enhances the understanding of legitimate behavior in democratic states and distinguishes it from illegitimate actions (Berzina & Soula, 2020).

A significant portion of research on interference by major powers through digital and communication technologies has focused on Russia, highlighting disinformation campaigns, hacking attacks on institutions, and individuals (Dawson & Innes, 2019; Baumann, 2020; Lemke & Habegger, 2022). Foreign powers typically influence elections through covert group agreements and by exploiting systemic and institutional vulnerabilities in digital communication technologies. Foreign actors have various methods at their disposal for using digital communication technologies to interfere in democratic elections. Tenove et al. (2018) identify four key digital techniques: hacking attacks on systems and databases, mass disinformation and propaganda campaigns, micro-targeted manipulation, and trolling. Social media platforms are particularly susceptible to trolling due to their design, which maximizes human engagement and advertising revenue. Disinformation campaigns, a tactic dating back to the Soviet era, have become widespread with the rise of social media platforms. Cyberattacks began notably around 2007, when Russia launched a campaign against Estonia (Authoritarian Interference Tracker, The Alliance for Securing Democracy). In October 2019, the European Parliament adopted a resolution condemning foreign interference in elections and disinformation processes within national and European democratic contexts (European Parliament, 2019).

In addition to interfering in elections, foreign actors can limit public discussions, restrict opportunities for public participation, and interfere with various rules and regulations. Under the influence of major powers, domestic actors often collaborate with foreign entities, employing diverse techniques to meddle in elections. Different countries exhibit varying levels of vulnerability to digital interference (Johnson, 2019; Helmus et al., 2020), influenced by factors such as deficits in digital literacy, poor design and policies of social media platforms, high levels of political polarization, and the absence or inadequacy of international laws and procedures for addressing cyber threats. Although numerous measures exist to counter digital interference, proven solutions remain elusive. Foreign actors can significantly exploit digital technologies to promote false or misleading information, deliberately suppress or filter information flows, and target social media platforms and various media institutions. Their aim is to generate polarizing and disapproving messages, thereby undermining processes for creating legitimate collective solutions to social problems.

Some research explores the impact of sharp power as a new form of hidden influence by Russia, China, and other authoritarian countries on the political systems of democracies (Greene, 2020). Sharp power has enabled Russia and China to expand their global influence. While not a new tactic, sharp power's effectiveness has increased due to technological advancements and the rising power of these great nations. Sharp power instruments were used in the former Soviet Union and Warsaw Pact countries at a lower level in the late 1990s and early 2000s (Gvosdev, 2019). According to Berzina and Soula (2020), the term "sharp power" is too limited to describe the array of covert actions by authoritarian states to destabilize democracies. It does not include economic interference or aggressive investments in technological infrastructure. Brutger et al. (2023) find that even economic policies not explicitly designed to influence elections can be perceived by the public as election interference.

According to the latest research, 49% of technological experts believe that technological development will weaken key aspects of democracy and democratic representation in the future, while only 33% believe it will strengthen them (Anderson & Raine, 2020). The rise of disinformation threats and deepfakes poses significant information risks for building democracy. It is crucial to explore why democracies have recently become vulnerable to brute force, major power intervention, and the role of social media in this process (DiPaola and Calo, 2024). This exploration aligns with the need for new tools and policies that democracies can use to defend themselves against information warfare campaigns (Collison, 2017; Čižik & Masariková, 2018). The effectiveness of these tools can vary by country, making research on brute force challenging. Understanding how democracies can respond to such threats is essential (Glazunova et al., 2023). Meanwhile, social media platforms must take greater responsibility for their actions. From a political science perspective, it is crucial to develop measures to understand cyber technologies and threats from information warfare.

Further, it is essential to develop analytical and forecasting mechanisms that provide insights into the future global leadership roles of the Russian Federation and China. These authoritarian countries will have tools to spread propaganda and disinformation. As technological development advances, accounting for the altering dynamics of the world and achieving a global power balance will become increasingly challenging. It will be essential to determine when cyber interference constitutes a violation of international law, including the rules prohibiting interference in a country's internal affairs or sovereignty, and ensuring the protection of international human rights (Schmitt, 2021). It is highly unlikely that countries will agree to more detailed and binding international rules if current regimes benefit from constructive ambiguity (Keitner, 2020). Partisan election interference can significantly influence election outcomes and other areas (Levin, 2019). Several studies have mapped the aspects of electoral influence and the international impact of electoral interference on democratization and democracy (Levin, 2016; Birch & Muchlinski, 2018). Foreign actors exploit systemic and institutional vulnerabilities in digital communication technologies—such as deficits in digital literacy and personal data protection, polarized political cultures and media systems, problematic social media design and policies, insufficient electoral and legal regulations, as well as insufficient international standards and laws on cyber security (Tenove et al., 2018).

Polarization can be intensified by homophilic information communities that develop within fragmented and polarized media systems (Tenove et al., 2018). In the digital era, the regulatory processes of campaigns and elections should be reformulated. Governments must seek optimal tools to detect interference methods applied by major powers in democratic institutions. Controversy surrounds the best approach for regulating the Internet by state authorities. According to Dunne (2021), the restrictive regulation of Internet platforms in open societies sets a dangerous precedent and may encourage authoritarian regimes to continue censorship. Designing effective tools for detecting computer crime, social media abuse, and cyber interference is crucial (Drobotov et al., 2023; MacDermott et al., 2022). Many experts also criticize the weaknesses in international law regarding cyber activities (Ohlin, 2017). The Internet was designed to connect, but it is increasingly being used to divide (Turnbull, 2017).

8.2.2 State of the Art: Threats of Disinformation Structure Development to Elections, Trust, and Democratic Legitimacy

The Internet and information technology have amplified the reach of social media, transforming them into potent tools for many countries (Kakar et al. 2023). These modern media platforms are increasingly wielded to influence democratic processes and propagate targeted negative agendas. The Internet eliminates geographical barriers, enabling global actors to connect and interact seamlessly. Consequently, modern warfare has evolved into a high-tech arena where social media play a crucial role. The reach and accessibility of information have expanded, with new media forms addressing larger audiences, contributing to direct communication, and ensuring anonymity when required. These structural changes in political communication create significant challenges. Big data analytics and algorithmic management have emerged as new forms of societal management (Jarrahi et al., 2021).

The development of digital technologies has led to a significant increase in the use of disinformation in international politics (Bradshaw & Howard, 2018; Lanoszka, 2019). Disinformation is now a powerful tool for influencing electoral processes, with numerous negative impacts. It can sway voter choices and discourage electoral participation (Zimmermann & Kohring, 2020). Kalla and Broockman (2018) found that pre-election advertisements have little effect on candidate selection; however, such ads are effective in altering voter turnout and influencing votes for lesser-known candidates (Aggarwal et al., 2023). Digital techniques are also employed to blackmail, threaten, or harass candidates, aiming to deter them from elections and public debates. This includes attacks on private data, threats from trolls, and online harassment of vulnerable groups (Lumsden & Morgan, 2017). Technological advancements facilitate new methods for creating and spreading disinformation, leading to severe consequences for public and political discourse, impacting political integrity, accountability, and elections.

Many experts assert that the rise of disinformation structures in various countries is closely linked to issues within national politics and traditional media (Egelhofer et al., 2020; Al-Mazahra, 2024). Consequently, addressing disinformation processes demands rigorous investigation and heightened accountability (Landon-Murray et al., 2019).

Foreign interference in elections poses a significant threat to electoral justice, democratic legitimacy, and public trust. The dissemination of propaganda and disinformation greatly impacts the global power balance, highlighting the importance of examining technological advancements and their effects on political systems, especially in an international context (Royal & Napoli, 2022). Numerous vulnerabilities within the electoral ecosystem require new approaches and strategies for protection. These include implementing campaign finance regulations to prevent foreign influence, reducing disinformation's impact on democratic discourse, bolstering state security agencies, and adopting measures to thwart computer hacking and privacy invasions (Dawood, 2021). To build resilience against disinformation, countries must systematically support critical thinking, raise awareness of disinformation issues, develop guidelines for effective fact communication and counterarguments, and establish networks of stakeholders committed to combating disinformation. Educational systems and policies are crucial in this process, promoting awareness and increasing trust (Popescu, 2020).

Although propaganda and various forms of influence have always been present in politics, strong technological development has disrupted the balance related to the use of false or misleading information. Disinformation will continue to be a very challenging and serious issue for countries. Hence, finding solutions with a global impact will be crucial. A lack of transparency is typical for today's large digital platforms, making accurate measurement of the social impact of disinformation almost impossible (Schia & Gjesvik, 2020; Sonmez et al., 2023). Despite the increase in tools to support election campaigns, many research studies confirm that little attention is paid to policies that would protect against information warfare campaigns and strategies (Tuttle, 2019). Conducting research in this area will be quite challenging, requiring an examination and understanding of how democracies might respond to these threats and identifying effective protective tools. There are also suggestions that social networks and other media enterprises should bear more responsibility, particularly in monitoring political advertisements to prevent the dissemination of false information in paid ads (Reisach, 2021).

Cyber operations have been increasingly assigned high strategic impact in many countries due to political fragmentation, intensifying international rivalry, complex inter-country relations, relative imbalances in military forces and capabilities, and socio-technological vulnerabilities (Ducheine & Pijpers, 2021). Recent conflicts, such as the war in Ukraine and the Israeli-Palestinian conflict, have made this trend more pronounced. Cyber tools, characterized by high flexibility, adaptability, speed, scalability, and limited potential for escalation, are relatively available and accessible (Linkov & Kott, 2019). These operations are effective, and their tools can enhance and transform traditional grey zone activities, including espionage. Therefore, it is essential to study these aspects and gather empirical evidence showing that disinformation has a strong destabilizing effect on democracies, whether examined at the microeconomic, mesoeconomic, or macroeconomic level (Schünemann, 2022).

Effective government intervention can also prevent the spread of disinformation and politically polarizing content. This requires a comprehensive understanding of the power and reach of cyber technologies and information warfare (Dunn & Wenger, 2022). Schmitt (2021) confirms that cyber operations related to elections have spurred the development of international laws against foreign electoral interventions through digital means, imposing obligations on states to end hostile cyber interference in electoral processes in accordance with international human rights law. Keitner (2020) explores the possibilities and limitations of international law in regulating mutual electoral influences between countries, noting that countries are unlikely to agree on more detailed, binding international rules and will therefore prefer to benefit from constructive uncertainty.

The importance of regulatory processes for addressing the problem of disinformation structures and their destabilizing effects on democracy is clear. However, regulatory measures must not restrict information freedom (Vese, 2022). The development of research into disinformation processes, with a reflective and critical approach, is essential. Civil society's involvement in fact-checking regimes, holding relevant actors accountable, is also important. Disinformation campaigns weaken institutions and support democratic decline (Bennett & Livingston, 2018). For institutions responsible for electoral integrity, past experiences are insufficient, as new disinformation tools bring greater threats and risks. The phenomenon of deepfakes highlights this challenge, as manipulated audio and video will make it harder to distinguish truth from lies, posing serious threats to future electoral processes (Diakopoulos & Johnson, 2021; Łabuz & Nehring, 2024).

Pressure for greater transparency on social media is evident in many countries. Social media platforms should disclose how they combat disinformation threats and create mechanisms to increase public trust (Reuter et al., 2019). The lack of transparency in social media and information security policies has weakened global efforts against disinformation. Disinformation campaigns affect not only democratic institutions, but also corporate and religious institutions, as well as individuals. Various democratic processes in different countries have specific issues that require urgent solutions, justifying increased interest in investigating the impact of disinformation and unfair practices during election periods (Santos, 2020).

The cited research studies examine various aspects of security strategies, threats, and policies, revealing a diverse array of content and processes. However, they lack an investigation into the relationship between political party preferences and perceived threats from major powers' interference. Additionally, few studies emphasize increasing digital media literacy as an effective tool for mitigating security threats. Moreover, existing literature primarily focuses on how major authoritarian states undermine democracies through electoral interference. Our research goes beyond this sole focus by investigating how citizens perceive electoral interference by major democratic powers, including the European Union as an intergovernmental institution and collective actor. The issue may stem from a lack of relevant research and access to deeply structured data. This study addresses these gaps and offers an innovative research framework with surprising empirical findings.

8.3 Data and Methodology

8.3.1 Data

The dataset was collected through a unique survey conducted July 17–25, 2023. Geographically, it includes respondents from three countries: 1069 from the Czech Republic, 1001 from Poland and 1005 from Slovakia. The selection of respondents was performed by a professional agency specializing in this field, and the sample is statistically representative, including the following demographic indicators studied: age, sex and education level.

The first question explored in the survey was: "To what extent do you think the following countries and international institutions should be the closest partner of the examined country?" Respondents selected their answers using a Likert scale from 1 to 5, where 1 meant absolutely satisfied, 2 meant rather satisfied, 3 meant neither satisfied nor dissatisfied, 4 meant rather dissatisfied, and 5 meant absolutely dissatisfied. Next, respondents assessed the degree of closeness of the explored countries using a four-point Likert scale. Here, 1 indicated that the country or institution should be a very close and important partner, while 4 suggested it should not be a partner. An additional option allowed respondents to indicate if they did not know. The countries and international institutions examined were the ones the respondents were from.

The second question explored was: "If elections to the parliament were held tomorrow and political parties ran separately, which party would you vote for?" Responses included specific political parties in the examined countries. In Czechia, the survey included the following political parties: ODS (Civic Democratic Party) (1), TOP09 (Tradition Responsibility Prosperity 09) (2), KDU-ČSL (Christian and Democratic

Union—Czechoslovak People's Party) (3), Piráti (Czech Pirate Party) (4), STAN (Mayors and Independents) (5), ANO (ANO 2011) (6), ČSSD (Czech Social Democratic Party) (7), KSČM (Communist Party of Bohemia and Moravia) (8), SPD (Freedom and Direct Democracy) (9), Trikolóra (Tricolor Citizens' Movement) (10), Zelení (Green Party) (11), Svobodní (Free Citizens' Party) (12), Přísaha (The Oath) (13), and PRO (Law, Respect, Expertise) (14). In Poland, the parties are Koalicja Obywatelska (Civic Coalition) (1), Konfederacja (Confederation Liberty and Independence) (2), Lewica (The Left) (3), Prawo i Sprawiedliwość (Law and Justice) / Zjednoczona Prawica (United Right) (4), and Trzecia Droga (Third Way—a coalition of PSL (Polish People's Party) and Polska 2050) (5). In Slovakia, the parties included are HLAS—SD (Voice—Social Democracy) (1), SMER—SD (Direction—Social Democracy) (2), Sloboda a Solidarita (Freedom and Solidarity) (3), Progresívne Slovensko (Progressive Slovakia) (4), SME RODINA (We Are Family) (5), OĽANO a priatelia + Kresťanská únia + ZA ĽUDÍ (Ordinary People and Independent Personalities and Friends + Christian Union + For the People) (6), REPUBLIKA (Republic) (7), KDH (Christian Democratic Movement) (8), Aliancia—Szövetség (Alliance) (9), SNS (Slovak National Party) (10), Kotlebovci—ĽS Naše Slovensko (Kotlebists—People's Party Our Slovakia) (11), Maďarské fórum (Hungarian Forum), Občianski demokrati Slovenska (Civic Democrats of Slovakia), Za regióny (For Regions), Rómska koalícia (Roma Coalition), Demokratická strana (Democratic Party) (12), Demokrati (Democrats) (13), Modrí (Blues), and Most—Híd (Bridge) (14).

For Czechia and Slovakia, the aggregation level was set at the level of political parties; for Poland, it was set at the level of political coalitions. There were three reasons for this. First, the level of fragmentation: The party system in Poland is much more fragmented than in Czechia and Slovakia. The five major coalitions included no less than 25 political parties and movements, almost twice as many as in Czechia and Slovakia. Secondly, while a considerable share of individual political parties in Poland did not specify their inclinations or antagonisms towards the analyzed major powers, the political coalitions did. Our research choice made Poland more directly comparable to Czechia and Slovakia. And finally, practical reasons: the practical limitations related to the overall scope of our survey in Poland did not allow us to include all the individual parties due to their number.

The third question explored in the survey was: "Do you think that the upcoming elections to the parliament will be influenced by the interference of the major powers?" Responses are measured on a four-point Likert scale, where the lowest level indicated the highest probability of interference.

8.3.2 Methodology

The main methodological technique for the analytical processing is regression analysis (Galton, 1989). The regression models are examined using logistic regression. Partnership importance is evaluated via a four-level Likert scale, where lower values indicate a more important relationship. It is important to note that if a certain category of the observed indicator is missing in the tables representing the analytical outcomes, it means there were no responses from any of the surveyed respondents, making it impossible to provide an outcome for such a case. The study explores the three countries: Czechia (CZ), Poland (PL), and Slovakia (SK).

The subsequent research questions (RQs) investigated in this study are:

- RQ1: How does sex influence the perception of partnerships among the explored countries?
- RQ2: How does age influence the perception of partnerships among the explored countries?
- RQ3: How do education levels influence the perception of partnerships among the explored countries?
- RQ4: How does support for a particular political party influence the perception of major power interference in elections in the explored countries?

8.4 Analysis and Results

The analytical processing is divided into two sections. The first section focuses on socio-demographic indicators, and the second section examines the intersection of supported political parties and foreign interference in political situations. Firstly, we investigate sex as a socio-demographic indicator. Table 8.1 demonstrates the estimated values obtained through regression models for sex. This analysis will help determine whether sex has a significant influence on the perception of partnerships among the explored countries.

Table 8.1 Regression model coefficients for sex

Country	*Observed area*	*Value*	*Constant value*	*Partnership importance*		
				2	*3*	*4*
CZ	altogether	estimation	0.9326	1.5817	1.5145	1.0722
		significance	$2.44 . 10^{-1}$	$5.24 . 10^{-5}$	$8.56 . 10^{-3}$	$7.71 . 10^{-1}$
CZ	PL	estimation	0.7851	1.8464	2.2228	1.6790
		significance	$1.54 . 10^{-2}$	$5.41 . 10^{-5}$	$7.69 . 10^{-5}$	$8.39 . 10^{-2}$
CZ	SK	estimation	1.0285	1.4958	0.8047	0.5186
		significance	$7.08 . 10^{-1}$	$3.72 . 10^{-2}$	$4.47 . 10^{-1}$	$1.39 . 10^{-1}$
PL	altogether	estimation	0.8637	1.1983	1.3542	1.6281
		significance	$2.28 . 10^{-2}$	$8.48 . 10^{-2}$	$3.51 . 10^{-2}$	$4.23 . 10^{-2}$
PL	CZ	estimation	0.8095	1.1747	1.6537	1.6471
		significance	$1.95 . 10^{-2}$	$2.64 . 10^{-1}$	$8.35 . 10^{-3}$	$2.04 . 10^{-1}$
PL	SK	estimation	0.9234	1.2414	1.0396	1.5703
		significance	$3.84 . 10^{-1}$	$1.60 . 10^{-1}$	$8.61 . 10^{-1}$	$1.39 . 10^{-1}$
SK	altogether	estimation	0.9081	1.2127	1.6574	1.1929
		significance	$1.10 . 10^{-1}$	$7.46 . 10^{-2}$	$4.60 . 10^{-4}$	$4.60 . 10^{-1}$
SK	CZ	estimation	0.9863	0.7066	1.3340	0.6970
		significance	$8.53 . 10^{-1}$	$4.53 . 10^{-2}$	$2.06 . 10^{-1}$	$3.65 . 10^{-1}$
SK	PL	estimation	0.7725	1.8010	2.1060	1.8120
		significance	$1.33 . 10^{-2}$	$1.17 . 10^{-3}$	$1.39 . 10^{-4}$	$5.57 . 10^{-2}$

Source: Authors' own elaboration

Each country is observed through the perspectives of the other explored countries and the overall view as seen in Table 8.1. Czechia is perceived to be 58.17% more important by females than the remaining values, which are 51.45% and 7.22%. This means that the lesser the importance of the partnership, the lower the odds of its becoming. In Poland, Czechia is perceived by females as less important than in Slovakia. The odds of being of lower importance are higher by 84.64% for females. There are also increases by 122.28% and 67.90% for females than for males to the lower importance levels. The lesser the importance level, the higher the odds are assigned to the respondents from Poland than the ones from Slovakia. This is the most unimportant partnership observed.

Secondly, Poland is assigned a rising tendency of the partnership importance levels. There is a 19.83% higher probability for females that Poland is perceived as less important. The odds of becoming less important are increased to 35.42% and 62.81% for females. In Czechia, a partnership with Poland is perceived as less important than in Slovakia, as there are lower odds of being less important in general. The odds for the second and fourth levels are similar, but the third level is considerably more favorable for Slovakia. Slovakia also demonstrates a strong tendency to fluctuate, as it is perceived at the second level similarly to Poland with odds higher by 21.27%, but the remaining values of 65.74% and 19.29%, signifying increases too, behave variously. From a country perspective, the partnership with Slovakia is seen as very important by the respondents in the Czech Republic. There has been a 29.34% decrease in the number of respondents who would consider it to be at the highest level of importance and a 20.30% decrease in the number of respondents who would consider it to be at the lowest level of importance. The respondents in Poland clearly see the partnership with Slovakia as unimportant. There is an 80.10% chance that it would be seen as lower than the highest importance and a 110.60% and 81.20% increase in the odds for the lower level. This partnership is almost the least important of all the observed possibilities.

Table 8.2 presents the observed changes in the selected countries in comparison to the whole observed area, with a focus on sex.

Table 8.2 illustrates the variations among the countries. A discernible pattern emerges across all the countries included in the analysis. The lower the importance, the greater the change. A level below the highest importance is observed for the attitude towards Slovakia. In the case of Czechia, the value is lower, while in Poland, it is higher. For the penultimate importance level, the variations are more pronounced. The highest value is observed for Czechia. This state is perceived in Slovakia at almost a half level, while in Poland, it is perceived in the opposite manner through an increase of almost a half. The lowest level of importance is maintained in a

Table 8.2 Changes in selected countries compared to whole area by sex

Country	*Observed area*	*Constant value (%)*	*Partnership importance (%)*		
			2	*3*	*4*
CZ	PL	84.18	116.74	146.77	156.59
CZ	SK	110.28	94.57	53.13	48.36
PL	CZ	93.72	98.03	122.12	101.16
PL	SK	106.91	103.60	76.77	96.45
SK	CZ	108.61	58.27	80.49	58.43
SK	PL	85.06	148.52	127.08	151.92

Source: Authors' own elaboration

manner analogous to the previously discussed case, and the attitude toward Slovakia is observed to be strikingly similar to that observed at the level below the highest importance.

Secondly, age is examined as a sociodemographic indicator. Age is shown in ten-year groups. Table 8.3 presents the estimated values derived from the regression models for age.

Firstly, a strong pattern can be seen in attitudes towards the Czech Republic. The younger the respondents, the less important they perceive partnership with the Czech Republic to be. The youngest category exhibits a mere 15.82% decrease in the probability of Czechia being rated as less important than the highest importance level. Furthermore, the remaining

Table 8.3 Regression model coefficients for age

Country	*Observed area*	*Age group*	*Value*	*Constant value*	*Partnership importance*		
					2	*3*	*4*
CZ	altogether	2	estimation	2.2716	0.8418	0.7894	0.7450
			significance	$7.59 . 10^{-10}$	$4.14 . 10^{-1}$	$3.76 . 10^{-1}$	$4.32 . 10^{-1}$
CZ	altogether	3	estimation	2.7284	0.7202	0.6951	0.7612
			significance	$1.10 . 10^{-14}$	$1.15 . 10^{-1}$	$1.68 . 10^{-1}$	$4.51 . 10^{-1}$
CZ	altogether	4	estimation	2.9630	0.5033	0.3608	0.2336
			significance	$2.84 . 10^{-17}$	$1.34 . 10^{-3}$	$4.10 . 10^{-4}$	$1.31 . 10^{-3}$
CZ	altogether	5	estimation	2.8642	0.3981	0.2408	0.0537
			significance	$3.53 . 10^{-16}$	$3.54 . 10^{-5}$	$7.57 . 10^{-6}$	$1.47 . 10^{-4}$
CZ	altogether	6	estimation	1.9877	0.1765	0.1041	0.0387
			significance	$4.59 . 10^{-7}$	$3.41 . 10^{-9}$	$1.39 . 10^{-6}$	$1.89 . 10^{-3}$
CZ	PL	2	estimation	3.0000	0.5482	0.5641	0.7083
			significance	$7.67 . 10^{-7}$	$3.93 . 10^{-2}$	$8.51 . 10^{-2}$	$4.75 . 10^{-1}$
CZ	PL	3	estimation	3.9258	0.4415	0.3527	0.6686
			significance	$2.24 . 10^{-10}$	$4.19 . 10^{-3}$	$1.91 . 10^{-3}$	$3.90 . 10^{-1}$
CZ	PL	4	estimation	3.7778	0.4059	0.2138	0.1324
			significance	$8.19 . 10^{-10}$	$1.81 . 10^{-3}$	$2.32 . 10^{-5}$	$1.85 . 10^{-3}$
CZ	PL	5	estimation	3.3703	0.3297	0.1484	0.0371
			significance	$2.95 . 10^{-8}$	$2.21 . 10^{-4}$	$2.35 . 10^{-6}$	$2.35 . 10^{-3}$
CZ	SK	2	estimation	1.9074	1.5293	1.3985	0.5244
			significance	$1.21 . 10^{-4}$	$2.56 . 10^{-1}$	$6.30 . 10^{-1}$	$3.24 . 10^{-1}$
CZ	SK	3	estimation	2.1296	1.3305	2.9747	0.5635
			significance	$4.60 . 10^{-6}$	$4.45 . 10^{-1}$	$8.99 . 10^{-2}$	$3.61 . 10^{-1}$
CZ	SK	4	estimation	2.5555	0.5218	1.3047	0.3913
			significance	$5.05 . 10^{-9}$	$1.16 . 10^{-1}$	$6.95 . 10^{-1}$	$1.50 . 10^{-1}$
CZ	SK	5	estimation	2.6111	0.4788	0.8939	0.0766
			significance	$2.01 . 10^{-9}$	$7.88 . 10^{-2}$	$8.74 . 10^{-1}$	$2.03 . 10^{-2}$

(*continued*)

Table 8.3 (continued)

Country	*Observed area*	*Age group*	*Value*	*Constant value*	*Partnership importance*		
					2	*3*	*4*
CZ	SK	6	estimation	2.9815	0.5590	0.6710	0.0671
			significance	$3.74 . 10^{-12}$	$1.44 . 10^{-1}$	$5.82 . 10^{-1}$	$1.46 . 10^{-2}$
PL	altogether	2	estimation	2.0492	0.9252	0.8751	0.8430
			significance	$4.36 . 10^{-6}$	$7.43 . 10^{-1}$	$6.33 . 10^{-1}$	$6.77 . 10^{-1}$
PL	altogether	3	estimation	2.6884	0.8215	0.6413	0.6426
			significance	$4.27 . 10^{-11}$	$3.92 . 10^{-1}$	$1.09 . 10^{-1}$	$2.78 . 10^{-1}$
PL	altogether	4	estimation	3.3441	0.6230	0.4640	0.2447
			significance	$1.31 . 10^{-16}$	$3.82 . 10^{-2}$	$5.98 . 10^{-3}$	$2.89 . 10^{-3}$
PL	altogether	5	estimation	3.1474	0.6090	0.4054	0.2889
			significance	$6.13 . 10^{-15}$	$3.17 . 10^{-2}$	$1.74 . 10^{-3}$	$7.07 . 10^{-3}$
PL	altogether	6	estimation	3.6884	0.8190	0.2805	0.2219
			significance	$1.54 . 10^{-19}$	$3.65 . 10^{-1}$	$1.96 . 10^{-5}$	$1.43 . 10^{-3}$
PL	CZ	2	estimation	1.3823	1.5976	1.2946	1.2660
			significance	$1.50 . 10^{-1}$	$1.60 . 10^{-1}$	$4.79 . 10^{-1}$	$7.23 . 10^{-1}$
PL	CZ	3	estimation	2.5588	0.8305	0.4937	0.6840
			significance	$3.40 . 10^{-6}$	$5.61 . 10^{-1}$	$5.49 . 10^{-2}$	$5.64 . 10^{-1}$
PL	CZ	4	estimation	3.2352	0.6826	0.4718	0.1546
			significance	$2.18 . 10^{-9}$	$2.25 . 10^{-1}$	$3.40 . 10^{-2}$	$3.55 . 10^{-2}$
PL	CZ	5	estimation	3.1176	0.5747	0.3039	0.4812
			significance	$7.97 . 10^{-9}$	$8.55 . 10^{-2}$	$1.89 . 10^{-3}$	$2.78 . 10^{-1}$
PL	CZ	6	estimation	3.2352	1.2106	0.3416	0.1546
			significance	$2.18 . 10^{-9}$	$5.26 . 10^{-1}$	$3.94 . 10^{-3}$	$3.55 . 10^{-2}$
PL	SK	2	estimation	2.8885	0.5482	0.6232	0.5934
			significance	$2.03 . 10^{-6}$	$7.99 . 10^{-2}$	$2.97 . 10^{-1}$	$3.21 . 10^{-1}$
PL	SK	3	estimation	2.8515	0.8037	0.9119	0.6012
			significance	$2.80 . 10^{-6}$	$5.10 . 10^{-1}$	$8.32 . 10^{-1}$	$3.33 . 10^{-1}$
PL	SK	4	estimation	3.4810	0.5626	0.4597	0.2873
			significance	$1.11 . 10^{-8}$	$8.37 . 10^{-2}$	$9.00 . 10^{-2}$	$3.08 . 10^{-2}$
PL	SK	5	estimation	3.1848	0.6411	0.5966	0.1794
			significance	$1.51 . 10^{-7}$	$1.81 . 10^{-1}$	$2.50 . 10^{-1}$	$9.72 . 10^{-3}$
PL	SK	6	estimation	4.2588	0.4990	0.2114	0.2348
			significance	$1.24 . 10^{-11}$	$3.36 . 10^{-2}$	$2.17 . 10^{-3}$	$1.18 . 10^{-2}$
SK	altogether	2	estimation	1.7806	0.9741	1.2254	1.8956
			significance	$2.91 . 10^{-5}$	$9.00 . 10^{-1}$	$4.19 . 10^{-1}$	$1.33 . 10^{-1}$
SK	altogether	3	estimation	2.7075	0.6291	0.7275	1.0620
			significance	$1.29 . 10^{-14}$	$2.29 . 10^{-2}$	$2.03 . 10^{-1}$	$8.89 . 10^{-1}$
SK	altogether	4	estimation	2.8660	0.5888	0.3912	0.4362
			significance	$2.23 . 10^{-16}$	$9.18 . 10^{-3}$	$5.50 . 10^{-4}$	$9.13 . 10^{-2}$
SK	altogether	5	estimation	2.7928	0.4140	0.2387	0.1790
			significance	$1.46 . 10^{-15}$	$3.71 . 10^{-5}$	$2.42 . 10^{-6}$	$5.98 . 10^{-3}$
SK	altogether	6	estimation	2.2806	0.2193	0.1860	0.1645

(*continued*)

Table 8.3 (continued)

Country	*Observed area*	*Age group*	*Value*	*Constant value*	*Partnership importance*		
					2	*3*	*4*
			significance	$4.83 . 10^{-10}$	$2.24 . 10^{-9}$	$1.12 . 10^{-6}$	$8.88 . 10^{-3}$
SK	CZ	2	estimation	1.3558	2.0897	1.2908	2.2161
			significance	$7.61 . 10^{-2}$	$5.06 . 10^{-2}$	$5.24 . 10^{-1}$	$3.40 . 10^{-1}$
SK	CZ	3	estimation	2.0846	0.9994	0.7595	2.1621
			significance	$3.51 . 10^{-6}$	$9.99 . 10^{-1}$	$4.93 . 10^{-1}$	$3.34 . 10^{-1}$
SK	CZ	4	estimation	2.3558	1.4150	0.4245	1.0630
			significance	$3.49 . 10^{-8}$	$3.40 . 10^{-1}$	$4.98 . 10^{-2}$	$9.43 . 10^{-1}$
SK	CZ	5	estimation	2.3558	0.8844	0.3538	0.4252
			significance	$3.49 . 10^{-8}$	$7.49 . 10^{-1}$	$2.25 . 10^{-2}$	$3.98 . 10^{-1}$
SK	CZ	6	estimation	3.1693	0.8414	0.3681	0.4742
			significance	$1.12 . 10^{-14}$	$6.41 . 10^{-1}$	$1.76 . 10^{-2}$	$4.20 . 10^{-1}$
SK	PL	2	estimation	2.8694	0.5161	0.8464	1.2198
			significance	$1.34 . 10^{-5}$	$2.82 . 10^{-2}$	$6.38 . 10^{-1}$	$7.04 . 10^{-1}$
SK	PL	3	estimation	4.3041	0.3753	0.5089	0.5421
			significance	$2.87 . 10^{-10}$	$7.60 . 10^{-1}$	$5.40 . 10^{-2}$	$2.57 . 10^{-1}$
SK	PL	4	estimation	4.1736	0.3133	0.2852	0.1997
			significance	$7.53 . 10^{-10}$	$9.01 . 10^{-5}$	$8.54 . 10^{-4}$	$1.30 . 10^{-2}$
SK	PL	5	estimation	3.9128	0.2408	0.1460	0.0852
			significance	$5.25 . 10^{-9}$	$3.52 . 10^{-6}$	$7.96 . 10^{-6}$	$3.73 . 10^{-3}$

Source: Authors' own elaboration

odds to the lowest importance levels are 21.06% and 25.50%. The 35–44 years category exhibits a similar pattern of behavior. A partial break occurs for the 45–54 years category, where there is a 49.67% decrease in the likelihood that the partnership with the Czech Republic would be considered less important than the highest importance. The remaining values indicate a further decline. The same trend is observed in the older age categories, where the 55–65 years category is assigned a decrease of 94.63%, indicating that respondents perceive this partnership as the least important. In the oldest age category, the decrease is even more pronounced, at 96.13%. From the perspective of respondents in Poland, the overall conclusion is supported by the relatively lower values assigned to the second importance level. In Slovakia, however, the situation is somewhat different, with the higher importance levels in the younger age categories being assigned increasing odds. This indicates that the younger the respondent, the less important it is to them to have a partnership with the Czech Republic. This is true up to the penultimate level, where the odds of decreasing importance are observed. The 35–44 age category exhibited the greatest increase, with a 197.47% change in the penultimate importance level. The older age categories exhibit a similar pattern to that observed in Poland.

Secondly, Poland and the Czech Republic share some similarities. One difference is that the decrease in the perceived importance of Poland is less pronounced than in the Czech Republic. This is evident from the two views. Firstly, the decreases in perceived importance as people age are less significant. Neither one is higher than 40%. It is also noteworthy that the highest decreases, belonging to the oldest age categories, are at the levels of 71.11–77.81%. Interestingly, a decrease in the perception of the partnership with Poland being of lower importance than at the highest level is only 18.10%. On the other hand, the differentiation between the explored countries is substantially different from the case of Czechia. While the respondents in Czechia exhibit a range of behaviors, the ones in Slovakia appear to follow a discernible pattern. The most striking behavior is observed in the youngest age category in Czechia, where the increases in considering the partnership with Poland less important are 59.76%, 29.46%, and 26.60%. At first glance, these values may appear modest, but when viewed in the context of the other values, they represent a strikingly contrasting trend. The remaining cases appear to behave in the way that has already been mentioned above. There is only a single exception, where the oldest age category shows an increase of 21.06% to demonstrate a lower level of importance than the highest level. It seems that respondents from Slovakia perceive the partnership with Poland to be more important. This is confirmed by the decreases in odds as respondents age and the partnership becomes less important. The highest decreases belong to the oldest age categories as previously mentioned.

Thirdly, Slovakia represents a country that behaves in a manner that is partially similar to Poland. The respondents view the partnership with Slovakia as being of greater importance in general, although there are two exceptions. This is due to the decreasing odds en route to rising age and lesser importance. The first exception is attributed to a markedly rising tendency among the youngest age category, where a mere 2.59% decline is observed at a level below the highest importance. This is followed by increases of 22.54% and 89.56% for the two remaining lower importance levels. The second exception is based on the 35–44 years category, where an opposite trend is observed. The lower the importance, the lower the decrease, and there is even a very slight increase of 6.20% for the lowest importance level. The remaining older age categories perform as in the previous cases. The attitude of the respondents in the Czech Republic is absolutely heterogeneous. The first three youngest age categories include considerable decreases for the third importance level. Further research is required to ascertain the origin of this behavior. Additionally, the youngest age category considers the partnership with Slovakia unimportant, as there are increases in the shift of its responses into the lower importance levels by 108.97%, 29.08%, and 121.61%. A similar pattern emerges in the 35–44 years category, which shows an increase of 116.21% to the lowest importance. The 45–54 years category is characterised by an increase of 41.50% to a lower level than the highest importance, followed by a decrease of 57.55%.

Table 8.4 illustrates the changes observed in the selected countries in comparison to the overall observed area, with regard to the age of respondents.

Table 8.4 illustrates the variations between the countries from the perspective of respondents from the countries under investigation. The perception of the Czech Republic held by respondents in Poland is considered to be consistent across all age categories. In contrast, respondents in Slovakia perceive it in a more differentiated manner.

Figure 8.1 illustrates the average changes by age as a socio-demographic indicator.

As Fig. 8.1 illustrates, there are striking similarities in the average changes observed across the individual categories examined within each age group. All the age categories under consideration exhibit distinctive characteristics within the countries studied. The darker the cell, the higher the average values of the regression model coefficients. The highest values are observed at the third importance level in the Czech Republic, as perceived by respondents in Slovakia, and vice versa at the second and fourth importance levels in Slovakia, as perceived by respondents in the Czech Republic.

Thirdly, education levels are subjected to scrutiny as a sociodemographic indicator. Subsequently, Table 8.5 presents the estimated values derived from the regression models for education levels.

As illustrated in Table 8.5, the education level represents a socio-demographic indicator that is considerably differentiated. Firstly, the partnership with the Czech Republic is perceived to be of great importance to the respondents in general. The higher the education level attained, the higher the perceived importance of the partnership. For instance, the two lowest

Table 8.4 Changes in the selected countries compared to the whole observed area by age

Country	*Observed area*	*Age group*	*Constant value (%)*	*Partnership importance (%)*		
				2	*3*	*4*
CZ	PL	2	132.06	65.12	71.46	95.08
CZ	PL	3	143.89	61.31	50.74	87.84
CZ	PL	4	127.50	80.65	59.26	56.65
CZ	PL	5	117.67	82.80	61.61	69.05
CZ	SK	2	83.97	181.66	177.17	70.39
CZ	SK	3	78.05	184.75	427.95	74.02
CZ	SK	4	86.25	103.67	361.63	167.48
CZ	SK	5	91.16	120.25	371.24	142.65
CZ	SK	6	150.00	316.69	644.60	173.39
PL	CZ	2	67.46	172.66	147.94	150.18
PL	CZ	3	95.18	101.10	76.98	106.44
PL	CZ	4	96.74	109.56	101.67	63.16
PL	CZ	5	99.05	94.37	74.96	166.54
PL	CZ	6	87.71	147.81	121.80	69.66
PL	SK	2	140.96	59.25	71.22	70.39
PL	SK	3	106.07	97.83	142.18	93.55
PL	SK	4	104.09	90.30	99.06	117.40
PL	SK	5	101.19	105.26	147.18	62.11
PL	SK	6	115.46	60.92	75.36	105.83
SK	CZ	2	76.15	214.53	105.34	116.91
SK	CZ	3	77.00	158.87	104.41	203.58
SK	CZ	4	82.20	240.31	108.52	243.71
SK	CZ	5	84.35	213.61	148.21	237.47
SK	CZ	6	138.97	383.77	197.88	288.32
SK	PL	2	161.15	52.98	69.07	64.35
SK	PL	3	158.97	59.66	69.95	51.05
SK	PL	4	145.62	53.21	72.91	45.78
SK	PL	5	140.10	58.17	61.18	47.59

Source: Authors' own elaboration

importance levels are assigned odds of a 91.22% and 94.47% decrease for respondents with a tertiary education. The distribution of respondents across the countries under consideration is very similar and follows the set trend. Secondly, attitudes towards Poland vary more according to the level of education than attitudes towards the Czech Republic. Nevertheless, there is also a rising trend with an increasing level of education. It is noteworthy that the importance level below the highest one stands with almost

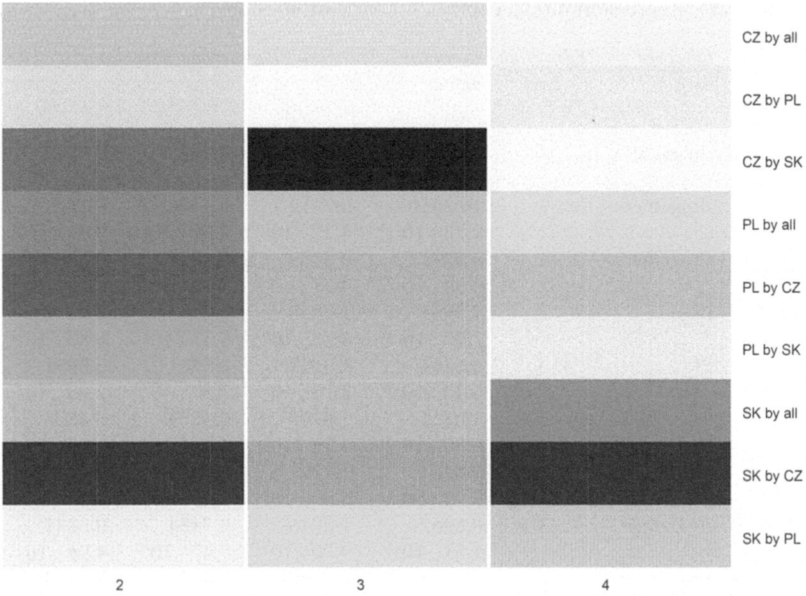

Fig. 8.1 Average changes by age category. (Source: Authors' own elaboration)

the same odds as the highest one itself from an overall perspective. A tendency to consider the partnership with Poland as important is only distorted by respondents with a secondary education level with diploma. These respondents are assigned a slight increase of odds at a level of 11.42 that they would consider it of lower importance. Thirdly, a situation in Slovakia is seen with the same tendency as in the previous two cases. However, some specificities are recorded here. Firstly, from a general perspective, respondents with tertiary education levels perceive the partnership with Slovakia to be significantly more important than those with other levels of education. This is evidenced by the fact that the odds of considering it to be of lesser importance are decreased by 71.63%, 87.00%, and 85.55%, respectively. Secondly, another specific situation can be observed in the particular countries that were examined. The respondents in the Czech Republic who have completed secondary education with a diploma are assigned an increase of 33.66% in the likelihood that they would consider the partnership with Slovakia to be of lesser importance. Additionally, the respondents with tertiary education consider the first

Table 8.5 Regression model coefficients for education levels

Country	*Observed area*	*Education level*	*Constant value*	*Partnership importance*		
				2	*3*	*4*
CZ	altogether	2	11.6178	0.6149	0.4558	0.5021
			$7.49 . 10^{-43}$	$9.77 . 10^{-2}$	$1.38 . 10^{-2}$	$1.48 . 10^{-1}$
CZ	altogether	3	14.2650	0.7912	0.3176	0.3622
			$9.75 . 10^{-51}$	$4.17 . 10^{-1}$	$3.59 . 10^{-4}$	$3.44 . 10^{-2}$
CZ	altogether	4	6.0297	0.3159	0.0878	0.0553
			$2.94 . 10^{-22}$	$4.25 . 10^{-4}$	$7.35 . 10^{-8}$	$5.44 . 10^{-4}$
CZ	PL	2	4.2941	1.1865	0.8219	0.8150
			$1.97 . 10^{-14}$	$5.75 . 10^{-1}$	$5.57 . 10^{-1}$	$6.83 . 10^{-1}$
CZ	PL	3	6.6764	1.3409	0.5551	0.5991
			$5.47 . 10^{-25}$	$3.19 . 10^{-1}$	$7.39 . 10^{-2}$	$2.98 . 10^{-1}$
CZ	SK	3	1.0361	1.0998	0.4504	0.4826
			$6.89 . 10^{-1}$	$6.75 . 10^{-1}$	$1.75 . 10^{-2}$	$1.22 . 10^{-1}$
CZ	SK	4	0.8233	1.1299	0.3644	0.1735
			$3.92 . 10^{-2}$	$6.09 . 10^{-1}$	$9.94 . 10^{-3}$	$2.15 . 10^{-2}$
PL	altogether	3	0.8967	1.0046	0.7434	0.5841
			$1.51 . 10^{-1}$	$9.70 . 10^{-1}$	$7.47 . 10^{-2}$	$4.96 . 10^{-2}$
PL	altogether	4	0.7418	0.8193	0.6267	0.4172
			$1.85 . 10^{-4}$	$1.35 . 10^{-1}$	$1.13 . 10^{-2}$	$7.57 . 10^{-3}$
PL	CZ	3	0.7571	1.0974	0.5778	0.6604
			$8.13 . 10^{-3}$	$5.73 . 10^{-1}$	$1.63 . 10^{-2}$	$3.52 . 10^{-1}$
PL	CZ	4	0.5952	0.8524	0.6300	0.4200
			$4.38 . 10^{-6}$	$3.90 . 10^{-1}$	$5.63 . 10^{-2}$	$1.28 . 10^{-1}$
PL	SK	3	1.0823	0.9240	1.1142	0.4975
			$4.74 . 10^{-1}$	$6.63 . 10^{-1}$	$6.74 . 10^{-1}$	$4.57 . 10^{-2}$
PL	SK	4	0.9367	0.8007	0.7222	0.3695
			$5.68 . 10^{-1}$	$2.50 . 10^{-1}$	$2.67 . 10^{-1}$	$1.36 . 10^{-2}$
SK	altogether	2	15.2015	0.5501	0.3684	0.3618
			$3.00 . 10^{-47}$	$4.21 . 10^{-2}$	$1.16 . 10^{-3}$	$3.50 . 10^{-2}$
SK	altogether	3	14.7349	0.7897	0.3257	0.3507
			$3.98 . 10^{-46}$	$4.18 . 10^{-1}$	$2.95 . 10^{-4}$	$3.05 . 10^{-2}$
SK	altogether	4	5.7673	0.2837	0.1300	0.1445
			$8.05 . 10^{-19}$	$1.71 . 10^{-4}$	$2.36 . 10^{-7}$	$2.39 . 10^{-3}$
SK	CZ	3	0.7259	1.3366	0.5564	0.5166
			$1.97 . 10^{-4}$	$1.35 . 10^{-1}$	$3.14 . 10^{-2}$	$1.75 . 10^{-1}$
SK	CZ	4	0.5389	0.9970	0.5352	0.5798
			$5.60 . 10^{-11}$	$9.90 . 10^{-1}$	$4.23 . 10^{-2}$	$2.95 . 10^{-1}$
SK	PL	2	4.5000	1.1818	0.6667	0.6296
			$9.21 . 10^{-14}$	$5.87 . 10^{-1}$	$2.16 . 10^{-1}$	$3.70 . 10^{-1}$
SK	PL	3	6.9666	1.2462	0.5383	0.5981
			$2.72 . 10^{-23}$	$4.60 . 10^{-1}$	$5.18 . 10^{-2}$	$2.99 . 10^{-1}$

Source: Authors' own elaboration

two importance levels to be of similar importance. The remaining importance levels exhibit a similar trend as in the previous cases. The respondents in Poland consider the partnership with Slovakia to be of lesser importance than in the Czech Republic. The increases of 18.18 % and 24.62 % assigned to respondents with secondary education without and with a diploma for the importance level below the highest one are due to the aforementioned changes. The remaining levels exhibit a similar trend as previously observed.

Table 8.6 illustrates the changes observed in the selected countries in comparison to the overall observed area, in relation to education levels.

The data reveals significant variations among education levels in all the explored countries. This finding corroborates the results of the previous analysis, which indicated that the importance of partnerships shifts in response to changes in the level of electoral interference by major powers.

Figure 8.2 presents the average changes by education level as a socio-demographic indicator.

It is evident that the second importance level is assigned considerably higher average alteration than the lower importance levels, as illustrated by Fig. 8.2. From a geographical perspective, Czechia and Slovakia exhibit a greater degree of alteration in values than Poland. Conversely, the lowest

Table 8.6 Changes by education level in selected countries compared to the whole observed area

Country	*Observed area*	*Education level*	*Constant value (%)*	*Partnership importance (%)*		
				2	*3*	*4*
CZ	PL	2	36.96	192.98	180.33	162.32
CZ	PL	3	46.80	169.47	174.78	165.40
CZ	SK	3	8.92	178.87	98.82	96.11
CZ	SK	4	5.77	142.81	114.74	47.91
PL	CZ	3	84.43	109.24	77.72	113.06
PL	CZ	4	80.24	104.04	100.53	100.66
PL	SK	3	120.69	91.97	149.87	85.17
PL	SK	4	126.27	97.73	115.24	88.57
SK	CZ	3	4.77	242.95	151.04	142.79
SK	CZ	4	3.66	126.25	164.31	165.36
SK	PL	2	29.60	214.82	180.98	174.02
SK	PL	3	47.28	157.82	165.25	170.56

Source: Authors' own elaboration

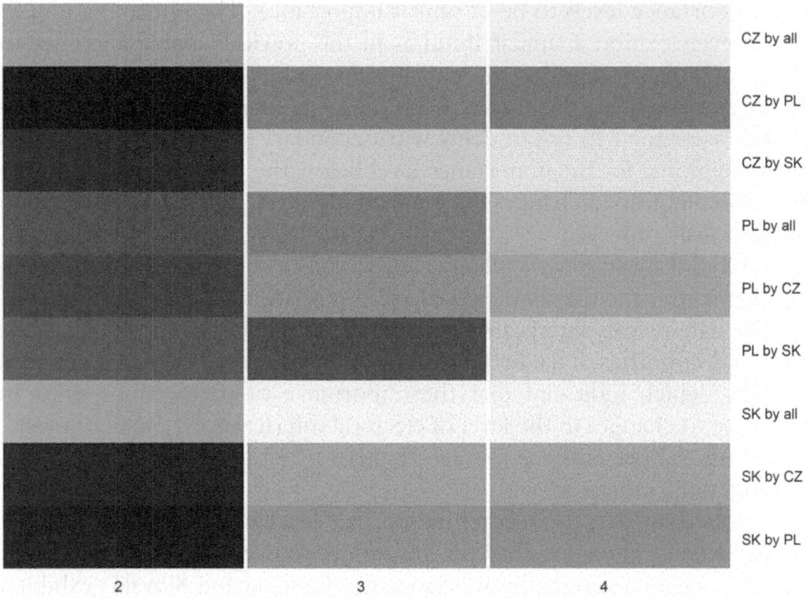

Fig. 8.2 Average changes by education level category. (Source: Authors' own elaboration)

importance level and the level above it are strikingly similar, indicating a higher degree of homogeneity in lower importance than in higher importance. A notable exception is observed in the perception of Poland by respondents in Slovakia.

The subsequent part of the analytical section is dedicated to an investigation of the role of major powers in influencing elections in the countries under consideration. This analysis examines allegiance to political parties in each country. For each political party, an odds ratio is calculated indicating the probability that a respondent who supports that party would consider the examined major power to be interfering in elections in the domestic country. The Tables that follow (8.7, 8.8, and 8.9) include, in addition to the assigned odds for each political party, the attitudes of these parties towards the major powers under examination. Tables 8.7, 8.8, and 8.9 include, in addition to the assigned odds for each political party, the attitudes of these political parties toward the examined major powers. A plus sign indicates a positive attitude toward the specified major power, a zero

Table 8.7 Czechia: Supported political parties/perception of foreign interference nexus

Political Party	CN	DE	RU	US	EU
1	9.0005 (-)	0.5789 (+)	Biased (-)	0.8125 (+)	0.8235 (+)
2	0.8889 (-)	3.1667 (+)	Biased (-)	4.0000 (+)	2.0238 (+)
3	1.0000 (-)	1.7273 (+)	Biased (-)	1.2308 (+)	1.2143 (+)
4	1.0000 (-)	2.8788 (+)	Biased (-)	2.3497 (+)	3.1032 (+)
5	0.6667 (-)	2.0152 (+)	Biased (-)	4.1026 (+)	4.0477 (+)
6	0.2993 (0)	6.1930 (+)	Biased (-)	8.8294 (+)	12.2857 (+)
7	0.7778 (+)	6.3334 (+)	Biased (0)	8.6154 (+)	8.5001 (+)
8	0.1481 (+)	Biased (-)	Biased (+)	Biased (-)	19.4286 (-)
9	0.1613 (+)	10.3637 (-)	Biased (+)	17.4768 (-)	10.4733 (-)
10	0.3333 (0)	5.1820 (-)	Biased (0)	Biased (-)	Biased (-)
11	Biased (-)	Biased (+)	Biased (-)	Biased (0)	Biased (+)
12	0.5555 (-)	8.6364 (-)	Biased (-)	Biased (0)	Biased (-)
13	0.2592 (0)	13.8183 (0)	Biased (-)	Biased (0)	12.1431 (-)
14	0.1111 (0)	11.2274 (-)	Biased (+)	Biased (-)	Biased (-)

Source: Authors' own elaboration

Table 8.8 Poland: Supported political parties/perception of foreign interference nexus

Political Party	CN	DE	RU	US	EU
1	2.2703 (0)	1.9758 (+)	6.2943 (-)	2.2368 (+)	2.9364 (+)
2	0.7419 (0)	6.8366 (+)	0.8121 (-)	3.8000 (+)	3.6103 (-)
3	1.0165 (0)	0.9784 (+)	1.0593 (-)	1.7882 (0)	0.9645 (+)
4	1.0088 (0)	8.6903 (0)	1.3866 (-)	0.7882 (0)	5.0612 (0)
5	0.5760 (0)	1.5182 (+)	0.6091 (-)	1.2853 (+)	4.7689 (+)

Source: Authors' own elaboration

Table 8.9 Slovakia: Supported political parties/perception of foreign interference nexus

Political Party	CN	DE	RU	US	EU
1	1.2942 (+)	2.7003 (+)	2.1536 (+)	7.4034 (0)	12.3355 (+)
2	0.2537 (+)	0.9332 (+)	0.3208 (+)	2.1604 (0)	1.5570 (+)
3	1.0817 (+)	0.7406 (+)	5.5748 (0)	0.4052 (+)	0.0946 (+)
4	3.6058 (0)	0.1481 (0)	7.4328 (0)	0.1801 (0)	0.0540 (+)
5	2.7816 (0)	0.8465 (+)	2.3219 (-)	0.4593 (+)	0.4053 (+)
6	0.7727 (0)	0.1852 (0)	1.0062 (+)	0.2927 (-)	0.1390 (-)
7	0.3512 (0)	1.1783 (+)	0.2322 (-)	Biased (+)	Biased (+)
8	1.1038 (0)	2.5922 (0)	0.7740 (0)	2.0264 (0)	0.5673 (+)
9	1.0302 (+)	2.5920 (-)	1.3933 (+)	Biased (-)	Biased (-)
10	0.3380 (0)	1.4073 (-)	0.1912 (+)	Biased (-)	0.9725 (-)
11	0.3566 (0)	1.4813 (-)	0.5224 (+)	Biased (-)	Biased (-)
12	Biased (0)	Biased (0)	Biased (0)	Biased (0)	Biased (+)
13	2.3177 (0)	0.1234 (+)	Biased (-)	0.0450 (+)	0.0811 (+)
14	1.5453 (0)	Biased (+)	0.9287 (-)	0.6753 (+)	Biased (+)

Source: Authors' own elaboration

signifies a neutral attitude, and a minus sign denotes a negative attitude. These attitudes are based on the declarations of the political parties in their political programs. The darker the highlighting, the more positive the attitude.

Firstly, the relations concerning the respondents in Czechia are scrutinized and illustrated in Table 8.7.

Table 8.7 presents the odds of respondents surveyed in the Czech Republic. The respondents supporting a political party with the highest odds of China interfering in forthcoming elections are those who support parties 1, 3, and 4. All other political parties are assigned lower odds, with party 14 having odds that are only 0.1111 times higher. In contrast, attitudes towards Germany are completely different. Party 1 is assigned the lowest odds of all that Germany would interfere in forthcoming elections. Conversely, 13 demonstrates the highest odds, which are 13.8183 times higher. Attitudes towards the United States are similarly aligned with Germany. Party 1 possesses the lowest odds, while 9 demonstrates the highest, which are 17.4768 times higher. The European Union is similarly situated, with 1 representing the lowest odds and 8 the highest, which are 19.4286 times higher.

Secondly, the attitudes of respondents in Poland are investigated and visualized in Table 8.8.

Table 8.8 presents the results of responses in Poland, indicating that differentiation is at its lowest level in this country.

Table 8.8 presents the results of the respondents' responses in Poland, indicating that differentiation is at its lowest level in this country. The supporters of 3 showed the highest odds of China's potential interference in the upcoming elections. The level is 1.0165 times higher than that of 1. Additionally, 4 is a very close approximation, with odds that are 1.0088 times higher than 1. The lowest odds are assigned to 5, with a level of 0.5760 times higher. The attitudes towards Germany are markedly distinct from those previously observed. The highest odds, assigned a level of 8.6903 times higher, are attributed to 4, while the lowest odds, assigned a level of 0.9784 times higher, are attributed to 3, which is a very close outcome to the reference party, Russia's behavior is also partially similar. The highest odds are assigned to option 4, with a level of 1.3866 times higher, while the lowest odds are assigned to option 5, with a level of 0.6091 times lower. The attitudes toward the United States demonstrate that the highest odds at a level of 3.8000 times higher are assigned to 2, while the lowest odds at a level of 0.7882 times higher are assigned to 4.

Finally, the attitude toward the European Union would be expected to influence the outcome of future elections at the lowest level of 0.9645 times higher, which is very close to 1. This is assigned to 3. The highest odds at a level of 5.0612 are assigned to 4.

Thirdly, the relations created according to respondents' attitudes in Slovakia are explored and presented in Table 8.9. Table 8.9 illustrates the attitudes of respondents supporting a particular political party in Slovakia toward the potential influence of major powers on the outcome of upcoming elections. Overall, the situation is more nuanced than in Poland, but less so than in the Czech Republic. Slovakia occupies a position between these two countries in terms of the perceived likelihood of external interference. Those who support option 4 are assigned the highest odds of 3.6058 that China will attempt to influence the outcome of the elections. Conversely, the lowest odds of 0.2537 are assigned to respondents supporting 2. A comparable distribution of odds is observed in the case of Germany, where the highest odds of 2.5922 belong to 8, followed very closely by 9, which is assigned odds of 2.5920. From an opposing perspective, the lowest odds of 0.1234 belong to 13. The attitudes toward Russia exhibit a markedly more differentiated pattern. The highest odds of 7.4328 are assigned to 4, while the lowest odds at a level of 0.1912 are assigned to 10. The United States demonstrates a situation in which the reference political party possesses the highest value of 7.4034. Subsequently, the highest odds of 2.1604 are assigned to 2, while the lowest odds of 0.0450 are assigned to 13. The attitudes toward the European Union are represented by the highest value of 12.3355 assigned to the reference political party. Only one political party, 2, is assigned odds of 1.5570 that are higher than for the reference party. The remaining political parties exhibit lower odds of the European Union's potential interference in upcoming elections. The minimum of 0.0540 is maintained by 4.

8.5 Discussion

Currently, there is a clear trend among European Union residents of decreasing interest in their politicians and political parties (Del-Castillo-Feito et al., 2022). Many European Union member states exhibit high levels of polarity in their political party systems, making it extremely difficult to form stable governments. Citizens struggle with corruption, economic and social instability, the migration crisis, and resistance to continued participation in the European community. According to Blanco-González et al. (2019), greater political engagement and a stronger relationship between politicians and the populace can enhance a country's legitimacy. Numerous studies confirm that changes in political party attitudes can immediately influence public political opinions, even if these new attitudes contradict previous public opinions (Slothuus & Bisgaard, 2021).

This study addressed four research questions as outlined in the Methodology section. The first research question examined sex as a socio-demographic indicator influencing the perceived partnerships of the explored countries. Among all the indicators examined, sex exhibited the most homogeneous behavior. It is noteworthy that female respondents perceived all partnerships as less important compared to male respondents. Czechia was viewed as significantly more important in Slovakia and Poland, while Slovakia was considered moderately important in Czechia, but unimportant in Poland. Conversely, Poland was perceived quite homogeneously by women in both respondent countries.

The importance of socio-demographic characteristics in examining relationships with political parties and electoral policies is confirmed by several studies (Furnham & Cheng, 2019; Huff & Tingley, 2015). The research framework on this issue is continuously expanding, highlighting the need to investigate a broader range of determinants. For instance, Häusermann and Kriesi (2015) explored the impact of socio-demographic characteristics on views about immigration, European Union integration, universalism, and welfare state chauvinism. They found that education directly affects economic liberalism, with income having an even stronger impact. Del-Castillo-Feito et al. (2022) also emphasize the importance of demographic factors, noting that pluralistic and divided societies are more prone to conflict and instability due to the complexities in constructing political systems. Age plays a crucial role in attitudes towards partnerships, as addressed by the second research question. Generally, older respondents perceive these partnerships as less important. In Slovakia, partnerships with Czechia are valued more than in Poland. In Poland, the perceived importance of partnerships decreases with age. In In Slovakia, a notable shift occurs for respondents aged 45–54, with younger respondents viewing the partnership as moderately important, while older respondents see it as less important. Respondents over 45 years in Slovakia consider the partnership highly important, while younger age groups view it as moderately important. In Czechia, younger respondents below 45 years view the partnership with Slovakia as unimportant, whereas older respondents assign it high importance. Polish respondents aged 25–34 see the partnership with Slovakia as partially unimportant, with older respondents valuing it more highly as age increases.

Thirdly, education levels are a socio-demographic indicator examined in the third research question. In all countries, more educated respondents place higher importance on partnerships. Partnership with Czechia is highly valued by tertiary-educated respondents, especially in Slovakia, while Polish respondents view it as moderately important. Partnership with Slovakia is viewed similarly to Czechia, with higher education levels correlating with greater importance. Tertiary-educated respondents in both countries assign the highest importance to these partnerships. Political parties strongly influence public opinion. According to Bullock (2020), individuals within each party respond homogeneously to changes in party positions, enabling parties to effectively shape opinions among their supporters. Coppock et al. (2018) confirm these homogeneous effects. Research is needed to uncover the psychological mechanisms behind party influence on individual opinions. The population is expected to evaluate political arguments and adapt their attitudes in an evolving reality.

Fourthly, the relations between supported political parties and perceptions of major power interference are examined based on the fourth research question. In Czechia, respondents supporting ODS (Civic Democratic Party) (1), KDU-ČSL (Christian and Democratic Union—Czechoslovak People's Party) (3), and Piráti (Czech Pirate Party) (4) have the highest odds of believing China will interfere in upcoming elections. Conversely, supporters of Přísaha (The Oath) (13), SPD (Freedom and Direct Democracy) (9), and KSČM (Communist Party of Bohemia and Moravia) (8) have the highest odds for expecting interference from Germany, the United States, and the European Union, respectively. The lowest odds of perceiving interference are assigned to supporters of PRO (Law, Respect, Expertise) (14) with respect to China and ODS (Civic Democratic Party) (1) with respect to Germany, the United States, and the European Union. In Poland, perceptions are more heterogeneous. Supporters of Trzecia Droga (Third Way—a coalition of PSL and Polski 2050) (5) have the lowest odds regarding perceived interference from China and Russia, Lewica (The Left) (3) regarding Germany and the European Union, and Prawo i Sprawiedliwość (Law and Justice) / Zjednoczona Prawica (United Right) (4) regarding the United States.

Conversely, Lewica (The Left) (3), Prawo i Sprawiedliwość (Law and Justice) / Zjednoczona Prawica (United Right) (4), and Koalicja Obywatelska (Civic Coalition) (1) have the highest odds with respect to China. Prawo i Sprawiedliwość (Law and Justice) / Zjednoczona Prawica (United Right) (4) supporters also perceive Germany, Russia, and the European Union as likely interferers, while Konfederacja (Confederation Liberty and Independence) (2) supporters view the United States at the highest odds. In Slovakia, perceptions are similarly diverse. Supporters of Progresívne Slovensko (Progressive Slovakia) (4) have the highest odds regarding Chinese and Russian interference, while SNS (Slovak National Party) (10) and Kotlebovci—ĽS Naše Slovensko (Kotlebists—People's Party Our Slovakia) (11) presume Germany, and SMER—SD (Direction—Social Democracy) (2) expects the United States and the European Union to interfere. The lowest odds are assigned to SMER—SD (Direction—Social Democracy) (2) regarding China, Demokrati (Democrats) (13) regarding Germany and the United States, SNS (Slovak National Party) (10) with respect to Russia, and Progresívne Slovensko (Progressive Slovakia) (4) regarding the European Union.

To increase public interest in politicians, the political situation, and involvement in public policies, optimal solutions must be sought. One potential approach is developing a system of political education and digital media literacy that aids understanding of the dynamic changes in social structures, relationships, and institutions, as framed by Bauman's liquid modernity concept (Bauman & Haugaard, 2008). In times of dynamic change, social structures become fragmented, and individuals may prioritize personal decisions and experiences over long-term commitments. This shift impacts attitudes toward political issues and the formation of relationships with political structures. Digital literacy is closely linked to the digital divide (Bach et al., 2018; Tirado-Morueta et al., 2018). Countries with a persistent or widening digital divide will experience low levels of digital literacy. This divide can significantly hinder public participation in policy-making and civic activities. As an individual's digital gap decreases, their participation in public policies tends to increase, aligning with the theory of technological determinism (Zara, 2021). Populations with lower levels of digital media literacy face greater difficulty recognizing the trustworthiness or origin of digital news and are more susceptible to manipulation. Thus, enhancing digital media literacy and civic education will significantly impact the vulnerability of populations to online disinformation, propaganda, and manipulation. There is no consensus on the best educational forms to mitigate these risks, although some successes have been noted (Maksl et al., 2017; Craft et al., 2017; Tully et al., 2022). Motivation and the capacity to evaluate content in highly fragmented environments are critical (Lazer et al., 2017). As Kahne and Bowyer (2017) assert, the identity and political awareness of today's youth are developed through social media in a highly polarized era, necessitating special attention to educational systems.

The Council of Europe's recommendations for creating a working group can serve as an inspiration to identify best practices in various domains essential for modern society. These include traditional intelligence literacy, social media forensics, algorithmic content design, the ethical implications of artificial intelligence, techniques for developing emotional skepticism, and cyber self-defense. Global knowledge agents employ cultures of uncertainty, fear, and irrational reaction to achieve their goals (Mölder & Shiraev, 2021). Therefore, creating effective monitoring and protection mechanisms is crucial. This involves systematically analyzing disinformation processes and manipulative interference, evaluating the application of European Union rules and regulations for personal data protection, implementing electoral guarantees, strengthening pluralistic media, supporting professional journalism, and ensuring optimal media literacy. Additionally, it is necessary to establish comprehensive legislation covering the financing of election campaigns to provide stricter guarantees (Irwin & van Holsteyn, 2021). Disinformation remains unregulated by international law, necessitating national legislation to regulate social and other media without infringing on freedom of speech and democratic principles (Nagasako, 2020).

8.6 Conclusion

The main goal of this study was to investigate specific aspects of foreign security policy through the perceived relationships between political party preferences and the influence of major powers on electoral processes in democratic countries. This perception was analyzed in relation to selected socio-demographic characteristics to identify a structural framework for developing a digital and media literacy system. Currently, no research studies have explored these relationships in the three neighboring countries selected, which can confirm the fundamental parameters of liquid modernity, such as fluidity, instability, fragmentation, risk, uncertainty, ethical and moral consequences, consumerism, and individualism.

The analytical outcomes demonstrate several interesting findings regarding the examined socio-demographic indicators. Age and education levels showed quite heterogeneous outcomes. In Czechia and Slovakia, this trend was stronger for education levels than in Poland. Altogether, female respondents considered the explored partnerships less important, indicating that sex was not as significant a differentiator as age and education level. Perceptions of major powers' interference in elections varied across the political parties. In Czechia, respondents supporting ODS (Civic Democratic Party) were assigned the highest odds of perceiving interference by China and the lowest by Germany, the United States, and the European Union. In Poland, the highest odds of interference by China and Germany were associated with supporters of PiS (Law and Justice), while supporters of KO (Civic Coalition) considered the United States the least likely to interfere. Conversely, KO supporters believed China had the highest odds of interference and the European Union the lowest. In Slovakia, no multiple occurrences were observed, but several cases were very close, indicating distinct perceptions of interference based on political alignment.

The technology sector has considerable potential for creating control mechanisms to prevent serious interference in democratic processes. Society should develop effective educational programs to help individuals recognize the differences between facts and fake news, trustworthy news sources, advertising biases, fairness, and more. A population demanding education and awareness constitutes the essential and best guarantee against foreign interference.

The study results are highly beneficial for security, educational, and social policymakers, national and international digital strategies, experts in information, media, and artificial intelligence literacy, regulators at the national and international levels, as well as experts dealing with the Digital Service Act and related legislation in the field of digital transformation. The outcomes will also benefit actors involved in the creation of disinformation strategies, those active in fact-checking, and many associations and institutions in the field of public policy.

References

Aggarwal, M., Allen, J., Coppock, A., et al. (2023). A 2 million-person, campaign-wide field experiment shows how digital advertising affects voter turnout. *Nature Human Behaviour, 7*(3), 332–341. https://doi.org/10.1038/s41562-022-01487-4

Aidt, T. S., Albornoz, F., & Hauk, E. (2021). Foreign influence and domestic policy. *Journal of Economic Literature, 59*(2), 426–487. https://doi.org/10.1257/jel.20201481

Albertazzi, D., & McDonnell, D. (2008). *Twenty-first century populism: The spectre of Western European democracy*. Palgrave Macmillan. https://doi.org/10.1057/9780230592100_14

Allison, G. (2017). *Destined for war: Can America and China escape Thucydides's trap?* Houghton Mifflin Harcourt.

Al-Mazahra, M. (2024). The circulation of fake news on social networks and its impact on the confidence of Jordanian journalists and media elites. *Journal of Media Studies and Research, 4*(13).

Anderson, J., & Raine, L. (2020). *Many tech experts say digital disruption will hurt democracy*. Pew Research Center. https://www.pewresearch.org/internet/2020/02/21/many-tech-experts-say-digitaldisruption-will-hurt-democracy/

Anheier, H. K., & Filip, A. (2021). The Dahrendorf quandary, crisis severity, and country performance. *Global Policy, 12*, 298–312. https://doi.org/10.1111/1758-5899.12901

Atkinson, W. (2008). Not all that was solid has melted into air (or liquid): A critique of Bauman on individualization and class in liquid modernity. *The Sociological Review, 56*(1), 1–17. https://doi.org/10.1111/j.1467-954X.2008.00774.x

Authoritarian Interference Tracker, The Alliance for Securing Democracy. https://securingdemocracy.gmfus.org/about-us/

Au-Yeung, R., & Tagirova, A. (2024). Beyond general elections: How could foreign actors influence the prime ministership? *International Journal, 79*(1), 149–159. https://doi.org/10.1177/00207020241234240

Bach, A. J., Wolfson, T., & Crowell, J. K. (2018). Poverty, literacy, and social transformation: An interdisciplinary exploration of the digital divide. *Journal of Media Literacy Education, 10*(1), 22–41. https://doi.org/10.23860/jmle-2018-10-1-2

Bauman, Z. (2000). *Liquid modernity*. Polity Press.

Bauman, Z., & Bordoni, C. (2016). *State of crisis*. Polity Press.

Bauman, Z., & Haugaard, M. (2008). Liquid modernity and power: A dialogue with Zygmunt Bauman. *Journal of Power, 1*(2), 111–130. https://doi.org/10.1080/17540290802227536

Baumann, M. (2020). Propaganda fights' and 'Disinformation Campaigns': The discourse on information warfare in Russia-West relations. *Contemporary Politics, 26*(3), 288–307. https://doi.org/10.1080/13569775.2020.1728612
Beck, U. (1996). Neo-nationalism or a Europe of individuals. *Theoria: A Journal of Social and Political Theory, 88*, 55–68. http://www.jstor.org/stable/41802672
Beck, U. (2018). *The metamorphosis of the world: How climate change is transforming our concept of the world.* Polity Press.
Bene, M., Petrekanics, M., & Bene, M. (2023). Facebook as a political marketing tool in an illiberal context. Mapping political advertising activity on Facebook during the 2019 Hungarian European Parliament and local election campaigns. *Journal of Political Marketing, 22*(3–4), 248–266. https://doi.org/10.4324/9781032664972-6
Bennett, W. L., & Livingston, S. (2018). The disinformation order: Disruptive communication and the decline of democratic institutions. *European Journal of Communication, 33*(2), 122–139. https://doi.org/10.1177/0267323118760317
Berzina, K., & Soula, E. (2020). *Conceptualizing foreign interference in Europe. Alliance for securing democracy.* German Marshall Fund of the United States, 19.
Birch, S., & Muchlinski, D. (2018). Electoral violence prevention: What works? *Democratization, 25*(3), 385–403. https://doi.org/10.1080/13510347.2017.1365841
Blanco-González, A., Payne, G., & Prado-Román, C. (2019). Political variables and state legitimacy. *UNIE Business Research, 8*(2), 123–133. https://doi.org/10.3926/hdbr.243
Bogaards, M. (2018). De-democratization in Hungary: Diffusely defective democracy. *Democratization, 25*(8), 1481–1499. https://doi.org/10.1080/13510347.2018.1485015
Bradshaw, S., & Howard, P. N. (2018). The global organization of social media disinformation campaigns. *Journal of International Affairs, 71*(1.5), 23–32. https://ora.ox.ac.uk/objects/uuid:67cd8a98-8b3c-45c0-b07f5e25b25ea67a/files/madb637c676fc7e1c0d8bd552a32751bc
Brutger, R., Chaudoin, S., & Kagan, M. (2023). Trade wars and election interference. *The Review of International Organizations, 18*, 1–25. https://doi.org/10.1007/s11558-022-09464-2
Bullock, J. G. (2020). Party cues. *The Oxford Handbook of Electoral Persuasion*, 129–150. Oxford University Press. https://doi.org/10.1093/oxfordhb/9780190860806.013.2
Caldwell, M., & Henry, P. C. (2020). The continuing significance of social structure in liquid modernity. *Marketing Theory, 20*(4), 547–572. https://doi.org/10.1177/1470593120948111

Cancela, P., & Dubied, A. (2022). Stay strong, get perspective, or give up: Role negotiation in small-scale investigative journalism. *Journalism Studies, 23*(9), 1056–1076. https://doi.org/10.1080/1461670x.2022.2067585

Čižik, T., & Masariková, M. (2018). Cultural identity as tool of Russian information warfare: Examples from Slovakia. *Science & Military Journal, 1.*

Clemens, E. S. (2021). The constitution of citizens: Political theories of nonprofit organizations. In *The nature of the nonprofit sector* (pp. 261–275). Routledge.

Collison, C. (2017). *Russia's information war: Old strategies, new tools.* Working Draft. https://jsis.washington.edu/ellisoncenter/wp-content/uploads/sites/13/2017/05/collison_chris_Russia%E2%80%99s-Information-War-Old-Strategies-New-Tools-How-Russia-Built-an-Information-Warfare-Strategy.pdf

Coppock, A., Ekins, E., & Kirby, D. (2018). The long-lasting effects of newspaper op-eds on public opinion. *Quarterly Journal of Political Science, 13*(1), 59–87. https://doi.org/10.1561/100.00016112

Craft, S., Ashley, S., & Maksl, A. (2017). News media literacy and conspiracy theory endorsement. *Communication and the Public, 2*(4), 388–401. https://doi.org/10.1177/2057047317725539

Dahrendorf, R. (2003). Making sense of the EU: The challenge for democracy. *Journal of Democracy, 14*(4), 101–114. https://www.journalofdemocracy.org/articles/making-sense-of-the-eu-the-challenge-for-democracy/

Dawood, Y. (2021). Combatting foreign election interference: Canada's electoral ecosystem approach to disinformation and cyber threats. *Election Law Journal: Rules, Politics, and Policy, 20*(1), 10–31. https://doi.org/10.2139/ssrn.4416565

Dawson, A., & Innes, M. (2019). How Russia's Internet research agency built its disinformation campaign. *The Political Quarterly, 90*(2), 245–256. https://doi.org/10.1111/1467-923x.12690

Deibert, R. J. (2022). Subversion Inc: The age of private espionage. *Journal of Democracy, 33*(2), 28–44. https://doi.org/10.1353/jod.2022.0016

Del-Castillo-Feito, C., Cachón-Rodríguez, G., & Paz-Gil, I. (2022). Political disaffection, sociodemographic, and psychographic variables as state legitimacy determinants in the European Union. *American Behavioral Scientist, 66*(1), 86–105. https://doi.org/10.1177/0002764220981116

Deuze, M. (2008). The changing context of news work: Liquid journalism for a monitorial citizenry. *International Journal of Communication, 2*, 18.

Diakopoulos, N., & Johnson, D. (2021). Anticipating and addressing the ethical implications of deepfakes in the context of elections. *New Media & Society, 23*(7), 2072–2098. https://doi.org/10.2139/ssrn.3474183

DiPaola, D., & Calo, R. (2024). Socio-digital vulnerability. https://doi.org/10.2139/ssrn.4686874

Douglass, J. A. (2021). *Neo-nationalism and universities: Populists, autocrats and the future of higher education*. Johns Hopkins University Press.

Drobotov, S., Pertsev, R., Hrab, M., Fedytnyk, V., Moroz, S., & Kikalishvili, M. (2023). Forensic research of the computer tools and systems in the fight against cybercrime. *Journal of Information Technology Management, 15*(1), 135–162.

Ducheine, P. A., & Pijpers, P. B. (2021). The notion of cyber operations. In *Research handbook on international law and cyberspace* (pp. 272–296). Edward Elgar Publishing.

Dunn, C. M., & Wenger, A. (2022). *Cyber security politics: Socio-technological transformations and political fragmentation*. Taylor & Francis.

Dunne, N. (2021). Platforms as regulators. *Journal of Antitrust Enforcement, 9*(2), 244–269. https://doi.org/10.2139/ssrn.3665007

Egelhofer, J. L., Aaldering, L., Eberl, J. M., Galyga, S., & Lecheler, S. (2020). From novelty to normalization? How journalists use the term "fake news" in their reporting. *Journalism Studies, 21*(10), 1323–1343. https://doi.org/10.1080/1461670x.2020.1745667

Elliott, A., & Lemert, C. (2021). *The new individualism: The emotional costs of globalization revisited*. Routledge.

Ercan, S. A., & Hendriks, C. M. (2013). The democratic challenges and potential of localism: Insights from deliberative democracy. *Policy Studies, 34*(4), 422–440. https://doi.org/10.1080/01442872.2013.822701

European Parliament. (2019). European Parliament resolution of 10 October 2019 on foreign electoral interference and disinformation in national and European democratic processes (2019/2810(RSP)). https://eur-lex.europa.eu/legal-content/EN/TXT/?uri=CELEX%3A52019IP0031%2801%29

Furnham, A., & Cheng, H. (2019). Personality traits and socio-demographic variables as predictors of political interest and voting behavior in a British cohort. *Journal of Individual Differences, 40*(2). https://doi.org/10.1027/1614-0001/a000283

Gabriel, Y., & Lang, T. (2018). *The unmanageable consumer*. Sage Publications.

Galton, F. (1989). Kinship and correlation. *Statistical Science, 4*(2), 81–86. https://doi.org/10.1214/ss/1177012581

Ganuza, E., & Font, J. (2020). Experts in government: What for? Ambiguities in public opinion towards technocracy. *Politics and Governance, 8*(4), 520–532. https://doi.org/10.17645/pag.v8i4.3206

Giugni, M., & Grasso, M. T. (2018). *Citizens and the crisis*. Palgrave Studies in European Political Sociology, Palgrave Macmillan. https://doi.org/10.1007/978-3-319-68960-9

Glazunova, S., Bruns, A., Hurcombe, E., Montaña-Niño, S. X., Coulibaly, S., & Obeid, A. K. (2023). Soft power, sharp power? Exploring RT's dual role in

Russia's diplomatic toolkit. *Information, Communication & Society, 26*(16), 3292–3317. https://doi.org/10.1080/1369118x.2022.2155485

Götz, E., & Merlen, C. (2019). Russia and the question of world order. *European Politics and Society, 20*(2), 133–153. https://doi.org/10.4324/9780367853419

Greene, E. (2020). *Sharp power: How foreign election interference is changing the global balance of power.* WWU Honors Program Senior Projects, 367. https://cedar.wwu.edu/wwu_honors/367

Gvosdev, N. K. (2019). Is Russia sabotaging democracy in the West? *Orbis, 63*(3), 321–333. https://doi.org/10.1016/j.orbis.2019.05.010

Häusermann, S., & Kriesi, H. (2015). What do voters want? Dimensions and configurations in individual-level preferences and party choice. In P. Beramendi, S. Häusermann, H. Kitschelt, & H. Kriesi (Eds.), *The politics of advanced capitalism* (pp. 202–230). Cambridge University Press. https://doi.org/10.1017/cbo9781316163245.009

Helmus, T. C., Marrone, J. V., Posard, M. N., et al. (2020). *Russian propaganda hits its mark: Experimentally testing the impact of Russian disinformation on U.S. audiences.* RAND Corporation. https://doi.org/10.7249/rra704-3

Hiltunen, M. (2021). Online political advertising and disinformation during elections: Regulatory framework in the EU and member states. *Helsinki Legal Studies Research Paper, 68.* https://doi.org/10.2139/ssrn.3897836

Holmes, S., & Krastev, I. (2018). Explaining Eastern Europe: Imitation and its discontents. *Journal of Democracy, 29*(3), 117–128. https://www.journalofdemocracy.org/articles/explaining-eastern-europe-imitation-and-its-discontents-2/

Huff, C., & Tingley, D. (2015). "Who are these people?" Evaluating the demographic characteristics and political preferences of MTurk survey respondents. *Research & Politics, 2*(3). https://doi.org/10.1177/2053168015604648

Irwin, G. A., & van Holsteyn, J. J. (2021). Keeping our feet dry: Impediments to foreign interference in elections in the Netherlands. *Election Law Journal: Rules, Politics, and Policy, 20*(1), 54–69.

Jarrahi, M. H., Newlands, G., Lee, M. K., Wolf, C. T., Kinder, E., & Sutherland, W. (2021). Algorithmic management in a work context. *Big Data & Society, 8*(2). https://doi.org/10.1177/20539517211020332

Johnson, D. E. W. (2019). Russian election interference and race-baiting. *Columbia Journal of Race and Law, 9*(2), 191–264.

Kagan, R. (2018). *The jungle grows back: America and our imperiled world.* Knopf.

Kahne, J., & Bowyer, B. (2017). Educating for democracy in a partisan age: Confronting the challenges of motivated reasoning and misinformation. *American Educational Research Journal, 54*(1), 3–34. https://doi.org/10.3102/0002831216679817

Kakar, R. A., Shahzad, S., & Babar, A. M. (2023). The impact of social media marketing dimensions: interaction, entertainment, customization, trendiness, and word of mouth on purchase intention with the mediating role of brand image and brand trust for smartphone users in Quetta city. *International Journal of Contemporary Issues in Social Sciences, 2*(4), 1249–1261.

Kalla, J. L., & Broockman, D. E. (2018). The minimal persuasive effects of campaign contact in general elections: Evidence from 49 field experiments. *American Political Science Review, 112*(1), 148–166. https://doi.org/10.1017/s0003055417000363

Kamarck, E. (2018). Malevolent soft power, AI, and the threat to democracy. https://www.brookings.edu/research/malevolent-soft-power-ai-and-the-threat-to-democracy/

Katz, R. S., & Mair, P. (1995). Changing models of party organization and party democracy: The emergence of the Cartel party. *Party Politics, 1*(1), 5–28. https://doi.org/10.1177/1354068895001001001

Katz, R. S., & Mair, P. (2009). The Cartel party thesis: A restatement. *Perspectives on Politics, 7*(4), 753–766. https://doi.org/10.1017/S1537592709991782

Keitner, C. (2020). Foreign election interference and international law. In D. Hollis & J. Ohlin (Eds.), *Election interference: When foreign powers target democratic institutions.* Oxford Academic.

Koc-Michalska, K., Lilleker, D. G., Michalski, T., Gibson, R., & Zajac, J. M. (2021). Facebook affordances and citizen engagement during elections: European political parties and their benefit from online strategies? *Journal of Information Technology & Politics, 18*(2), 180–193. https://doi.org/10.1080/19331681.2020.1837707

Kondratov, E., & Johansson-Nogués, E. (2023). Russia's hybrid interference campaigns in France, Germany and the UK: A challenge against trust in liberal democracies? *Geopolitics, 28*(5), 2169–2199. https://doi.org/10.1080/14650045.2022.2129012

Kriesi, H. (2018). Revisiting the populist challenge. *Politologický časopis—Czech Journal of Political Science, 25*(1), 5–27. https://doi.org/10.5817/PC2018-1-5

Łabuz, M., & Nehring, C. (2024). On the way to deep fake democracy? Deep fakes in election campaigns in 2023. *European Political Science.* https://doi.org/10.1057/s41304-024-00482-9

Lagna, A., & Lenglet, M. (2020). The dark side of liquidity: Shedding light on dark pools' marketing and market-making. *Consumption Markets & Culture, 23*(4), 390–406.

Landon-Murray, M., Mujkic, E., & Nussbaum, B. (2019). Disinformation in contemporary US foreign policy: Impacts and ethics in an era of fake news, social media, and artificial intelligence. *Public Integrity, 21*(5), 512–522. https://doi.org/10.1080/10999922.2019.1613832

Lanoszka, A. (2019). Disinformation in international politics. *European Journal of International Security, 4*(2), 227–248. https://doi.org/10.1017/eis.2019.6

Lazer, D., Baum, M., Grinberg, N., Friedland, L., Joseph, K., Hobbs, W., & Mattsson, C. (2017). Combating fake news: An agenda for research and action. https://shorensteincenter.org/combating-fake-news-agenda-for-research/

Lee, R. L. (2022). Time, space, and power in digital modernity: From liquid to solid control. *Time & Society, 31*(1), 69–87. https://doi.org/10.1177/0961463x211016781

Lemke, T., & Habegger, M. W. (2022). Foreign interference and social media networks: A relational approach to studying contemporary Russian disinformation. *Journal of Global Security Studies, 7*(2). https://doi.org/10.1093/jogss/ogac004

Levin, D. H. (2016). When the great power gets a vote: The effects of great power electoral interventions on election results. *International Studies Quarterly, 60*(2), 189–202. https://doi.org/10.1093/isq/sqv016

Levin, D. H. (2019). Partisan electoral interventions by the great powers: Introducing the PEIG Dataset. *Conflict Management and Peace Science, 36*(1), 88–106. https://doi.org/10.1177/0738894216661190

Linkov, I., & Kott, A. (2019). Fundamental concepts of cyber resilience: Introduction and overview. *Cyber Resilience of Systems and Networks.* https://doi.org/10.1007/978-3-319-77492-3_1

Lumsden, K., & Morgan, H. (2017). Media framing of trolling and online abuse: Silencing strategies, symbolic violence, and victim blaming. *Feminist Media Studies, 17*(6), 926–940. https://doi.org/10.1080/14680777.2017.1316755

Lury, C. (2020). *Consumer culture.* Polity Press.

MacDermott, Á., Motylinski, M., Iqbal, F., Stamp, K., Hussain, M., & Marrington, A. (2022). Using deep learning to detect social media 'trolls'. *Forensic Science International: Digital Investigation, 43*, 301446. https://doi.org/10.1016/j.fsidi.2022.301446

Maksl, A., Craft, S., Ashley, S., & Miller, D. (2017). The usefulness of a news media literacy measure in evaluating a news literacy curriculum. *Journalism & Mass Communication Educator, 72*(2), 228–241. https://doi.org/10.1177/1077695816651970

Mason, P. (2016). *PostCapitalism: A guide to our future.* Farrar.

McKay, S., & Tenove, C. (2021). Disinformation as a threat to deliberative democracy. *Political Research Quarterly, 74*(3), 703–717. https://doi.org/10.1177/1065912920938143

Mölder, H., & Shiraev, E. (2021). Global knowledge warfare, strategic imagination, uncertainty, and fear. In H. Mölder, V. Sazonov, A. Chochia, & T. Kerikmäe (Eds.), *The Russian federation in global knowledge warfare. Contributions to international relations.* Springer. https://doi.org/10.1007/978-3-030-73955-3_2

Mudde, C. (2021). Populism in Europe: An illiberal democratic response to undemocratic liberalism (The government and opposition/Leonard Schapiro lecture 2019). *Government and Opposition, 56*(4), 577–597. https://doi.org/10.1017/gov.2021.15

Mudde, C., & Kaltwasser, C. R. (2017). *Populism: A very short introduction.* Oxford University Press.

Müller, J. W. (2022). *The critical infrastructure of democracy.* Los Angeles Review of Books, 28 August. https://lareviewofbooks.org/short-takes/the-critical-infrastructure-of-democracy/

Nagasako, T. (2020). Global disinformation campaigns and legal challenges. *International Cybersecurity Law Review, 1*(1), 125–136.

Nicolaidis, K., & Youngs, R. (2023). Reversing the Gaze: Can the EU import democracy from others? *JCMS: Journal of Common Market Studies, 61*(6), 1605–1621. https://doi.org/10.1111/jcms.13477

Noor, U. A., & Niaz, B. (2023). Media and hybrid war: Political influence and disinformation. *Journal of Education and Humanities Research (JEHR), 15*(1), 57–64. http://journal.uob.edu.pk/journal/index.php/jehr/article/view/390

Norris, P., & Inglehart, R. (2019). *Cultural backlash: Trump, Brexit, and authoritarian populism.* Cambridge University Press.

O'Hara, K. (2020). The contradictions of digital modernity. *AI & Society, 35*(1), 197–208.

Ohlin, J. D. (2017). Is the concept of the person necessary for human rights? In *International legal personality* (pp. 437–478). Routledge.

Paternotte, D., & Verloo, M. (2021). De-democratization and the politics of knowledge: Unpacking the cultural Marxism narrative. *Social Politics: International Studies in Gender, State and Society, 28*(3), 556–578. https://doi.org/10.1093/sp/jxab025

Pérez-Curiel, C., & Rivas-de-Roca, R. (2023). Social cohesion in times of crisis: The role of communication for democracies—Editors' introduction. *Social Sciences, 12*(9), 491. https://doi.org/10.3390/socsci12090491

Pope, A. E. (2018). Cyber-securing our elections. *Journal of Cyber Policy, 3*(1), 24–38. https://doi.org/10.1080/23738871.2018.1473887

Popescu, M. M. (2020). Media Literacy tools in combating disinformation and fake news in social media. *Bulletin of the Transilvania University of Braşov, 13*(1), 103–112. https://doi.org/10.31926/but.ssl.2020.13.62.1.11

Radaelli, C. M., & O'Connor, K. (2009). How bureaucratic élites imagine Europe: Towards convergence of governance beliefs? *Journal of European Public Policy, 16*(7), 971–989. https://doi.org/10.1080/13501760903226641

Reisach, U. (2021). The responsibility of social media in times of societal and political manipulation. *European Journal of Operational Research, 291*(3), 906–917. https://doi.org/10.1016/j.ejor.2020.09.020

Reuter, C., Kaufhold, M. A., Schmid, S., Spielhofer, T., & Hahne, A. S. (2019). The impact of risk cultures: Citizens' perception of social media use in emergencies across Europe. *Technological Forecasting and Social Change, 148*(1), 1–17. https://doi.org/10.1016/j.techfore.2019.119724

Roitto, M., & Holmila, A. (2021). Liquid neutrality: Paradoxes of democracy in Finnish and Swedish NATO discussions? In *The Nordic economic, social and political model* (pp. 91–123). Routledge.

Royal, A., & Napoli, P. M. (2022). Platforms and the press: Regulatory interventions to address an imbalance of power. In *Digital platform regulation: Global perspectives on internet governance* (pp. 43–67). Springer International Publishing.

Sabri, S. (2021). Individuals and society in liquid modernity. *Qalaai Zanist Journal, 6*(3), 984–1002. https://doi.org/10.25212/lfu.qzj.6.3.35

Santos, G. F. (2020). Social media, disinformation, and regulation of the electoral process: a study based on 2018 Brazilian election experience. *Revista de Investigações Constitucionais, 7*(2), 429–449. https://doi.org/10.5380/rinc.v7i2.71057

Schia, N. N., & Gjesvik, L. (2020). Hacking democracy: Managing influence campaigns and disinformation in the digital age. *Journal of Cyber Policy, 5*(3), 413–428. https://doi.org/10.1080/23738871.2020.1820060

Schlipphak, B., & Treib, O. (2017). Playing the blame game on Brussels: The domestic political effects of EU interventions against democratic backsliding. *Journal of European Public Policy, 24*(3), 352–365. https://doi.org/10.1080/13501763.2016.1229359

Schmitt, M. N. (2021). Foreign cyber interference in elections. *eStockton Center for International Law, 97*, 1–27. https://digital-commons.usnwc.edu/cgi/viewcontent.cgi?article=2969&context=ils

Schünemann, W. J. (2022). A threat to democracies? An overview of theoretical approaches and empirical measurements for studying the effects of disinformation. *Cyber Security Politics*, 32–47.

Schwörer, J. (2021). *The growth of populism in the political mainstream: The contagion effect of populist messages on mainstream parties' communication.* Springer.

Sethi, M. G. U. (2007). Influencing elections: Grey zone warfare. In *Grey zone warfare in the Indian context* (pp. 1–176). Synergy.

Shaw, S. (2022). Authoritarian leadership: Is democracy in Peril? *Philosophy & Social Criticism, 48*(9), 1247–1276. https://doi.org/10.1177/01914537211072882

Shuker, P., & Topor, L. (2021). Russian influence campaigns against NATO in the Baltic region: Spread of chaos and divide et impera. In *The Russian federation in global knowledge warfare: Influence operations in Europe and its neighbourhood* (pp. 295–314). Springer.

Slothuus, R., & Bisgaard, M. (2021). How political parties shape public opinion in the real world. *American Journal of Political Science, 65*(4), 896–911. https://doi.org/10.1111/ajps.12550

Sonmez, B., Makarovs, K., & Allum, N. (2023). Public perception of scientists: Experimental evidence on the role of sociodemographic, partisan, and professional characteristics. *PLoS One, 18*(7), e0287572. https://doi.org/10.1371/journal.pone.0287572

Standing, G. (2021). *The precariat: The new dangerous class.* Bloomsbury Academic.

Tenove, C., Buffie, J., McKay, S., & Moscrop, D. (2018). *Digital threats to democratic elections: How foreign actors use digital techniques to undermine democracy.* Research Report, Centre for the Study of Democratic Institutions, University of British Columbia. https://doi.org/10.2139/ssrn.3235819

Tilly, C. (2023). Inequality, democratization, and de-democratization. *Sociological Theory, 21*(1), 37–43. http://www.jstor.org/stable/3108607

Tirado-Morueta, R., Aguaded-Gómez, J. I., & Hernando-Gómez, Á. (2018). The socio-demographic divide in Internet usage moderated by digital literacy support. *Technology in Society, 55*, 47–55.

Tomz, M., & Weeks, J. L. (2020). Public opinion and foreign electoral intervention. *American Political Science Review, 114*(3), 856–873. https://doi.org/10.1017/s0003055420000064

Tully, M., Maksl, A., Ashley, S., Vraga, E. K., & Craft, S. (2022). Defining and conceptualizing news literacy. *Journalism, 23*(8), 1589–1606. https://doi.org/10.1177/14648849211005888

Turnbull, M. (2017). Speech introducing the National Security Legislation Amendment (Espionage and Foreign Interference) Bill 2017. *December, 7*, 2017.

Tuttle, D. (2019). Campaigns of disinformation: Modern warfare, electoral interference, and canada's security environment. https://doi.org/10.2139/ssrn.3437117

Urbinati, N. (2016). Reflections on the meaning of the "Crisis of Democracy". *Democratic Theory, 3*(1), 6–31. https://doi.org/10.3167/dt.2016.030102

Vachudova, M. A. (2021). Populism, democracy, and party system change in Europe. *Annual Review of Political Science, 24*, 471–498. https://doi.org/10.1146/annurev-polisci-041719-102711

Vese, D. (2022). Governing fake news: The regulation of social media and the right to freedom of expression in the era of emergency. *European Journal of Risk Regulation, 13*(3), 477–513. https://doi.org/10.1017/err.2021.48

Vestri, G. (2015). La democracia líquida como alternativa a la política clásica? un estudio contextual. *Estudios de Deusto: revista de Derecho Público, 63*(1), 403–422.

Von der Leyen, U. (2019). *A union that strives for more: My agenda for Europe.* Political Guidelines for the Next European Commission 2019–2024.

Wall, M. (2017). Pop-up newsroom: Liquid journalism for the next generation. In *Global journalism education in the 21st century: Challenges and innovations* (pp. 327–342). University of Texas at Austin.

Wigell, M. (2019). Hybrid interference as a wedge strategy: A theory of external interference in liberal democracy. *International Affairs, 95*(2), 255–275. https://doi.org/10.1093/ia/iiz018

Zara, A. R. (2021). From the digital divide to public participation: Access, use, literacy, and perceived benefit in social and political participation. https://doi.org/10.2139/ssrn.4674296

Zielonka, J., & Rupnik, J. (2020). From revolution to "Counter-Revolution": Democracy in Central and Eastern Europe 30 years on. *Europe-Asia Studies, 72*(6), 1073–1099. https://doi.org/10.1080/09668136.2020.1784394

Zimmermann, F., & Kohring, M. (2020). Mistrust, disinforming news, and vote choice: A panel survey on the origins and consequences of believing disinformation in the 2017 German parliamentary election. *Political Communication, 37*(2), 215–237. https://doi.org/10.1080/10584609.2019.1686095

CHAPTER 9

Under What Conditions Could the Law Be an Effective Tool in Combating Disinformation?

Filip Cyuńczyk

9.1 Introduction

Previous discussions regarding the threats posed by systemic disinformation have clearly illustrated the multifaceted and interdisciplinary nature of the issue. These discussions encompass purely political science aspects, social issues such as the societal impacts of disinformation, and the examination of disinformation from the perspective of social engineering. They also include attempts to reconstruct the disinformation framework and analyse the tools to counteract it, taking into account systemic risks, such as the instrumental use of the law. The latter is of particular significance to me, as there remains a lack of widespread reflection on the problem of using the law, especially criminal and administrative legal norms, to combat disinformation. While public discourse has engaged in discussions concerning the extent of recourse to the law (whether existing or yet to be

F. Cyuńczyk (✉)
SWPS University, Warsaw, Poland
e-mail: fcyunczyk@swps.edu.pl

M. Wenzel, K. Stasiuk-Krajewska (eds.), *Disinformation in Central Europe*, https://doi.org/10.1007/978-3-031-76871-2_9

established), there is a noticeable absence of broader contemplation on the efficacy of addressing this issue through legal means. This may stem from a prevailing belief that the fragmentation of Western societies and their associated rapid individualisation is closely linked to the juridification of various aspects of life, which until the mid-twentieth century were governed not by established legal norms but by the phenomenon of social order.

Therefore, in this chapter, I aim to explore the relationship between the law and politics in the context of disinformation, particularly its effective counteraction within the system of constitutional liberal democracy. My primary objective is to identify the key challenges that may arise when applying strict and detailed normative regulations at the level of legal texts to counter the phenomena of disinformation. I acknowledge that these challenges may manifest both within the internal circulation of a specific political community and when faced with external threats. The latter situation occurs when there is an observable clash between two opposing ideological blocs. While in the first case, I believe it is impossible to draw a clear line between cynical disinformation efforts by participants in public discourse (within a specific political community) and minor violations of its principles, it is possible to delineate a division in the second case. Therefore, my considerations will primarily address scenarios where the law is used—as a tool by political actors—to combat external disinformation. This implies that specific narratives produced externally are targeted not against particular political forces or ideological blocs within an organisation but against the community as a whole: its social legitimacy, founding myths, and narratives consolidating its axiology.

In the context of contemporary constitutional liberal democracy in the Western world, of which Central Europe is a part, these factors include faith in the law as the most significant tool for limiting pure politics, the system of liberal constitutionalism built upon it, and, finally, the axiology derived from these assumptions.

In terms of faith in the law, the focus is primarily on its content. As noted by Russell Hardin, "the content of law is commonly quite specific, and it is often backed by powerful sanctions that make us comply with it" (2001, p. 62). Furthermore, regardless of whether a given law is analysed from a positivist or natural law perspective, the issue of its acceptance (legitimisation) always arises, which is inherently tied to the element of social order (2001, p. 80). This element underpins effective law, which in its content should pursue the Rawlsian goal of achieving mutual advantages resulting from coexistence within a community based on a social contract (which we know as the constitution) (Rawls, 1971). In other words, the element of social order is crucial for the internalisation of the norms contained in legal provisions by society. Faith, in this case, can manifest on two levels: top-down and bottom-up. The top-down perspective is nothing more than the assumption by law-makers that the law will be effectively internalised by society. The bottom-up perspective can be understood as the belief within society that the normative structure it advocates will enable the maintenance of order in an individualised reality. For this chapter, the first perspective is key, i.e., identifying the motivations that allow decision-makers (influencing the content of established and then applied law) to believe that their actions will be socially effective and align with bottom-up expectations.

This is an ambitious but crucial assumption for the proper functioning of the constitutional liberal democracy system. While on the doctrinal level, it may seem that this system is optimal from the perspective of the contemporary world (as it considers individual rights, the specific character of each person, and should be established from the bottom up, i.e., by free individuals), in everyday functioning, it is not free from paradoxes. Two of these paradoxes were discussed by Kevin Olson, who referred to the "Paradox of Founding" and the "Paradox of Dynamic Indeterminacy" (Olson, 2007). In the first case, it concerns the fact that "at first glance, a legal or political order cannot be democratically founded, at least not in a procedurally legitimate sense" (Olson, 2007, p. 331). This observation is related to the idea that it is impossible to constitute a political community without the existence of some pre-political form of unity within the same group. This translates into the legitimacy of the system, which from a procedural point of view, is impossible to achieve. However, it can be achieved through dynamic constitutionalism, where "a legitimate constitution need not be fully satisfactory from the start. A set of procedures could be exclusionary in the beginning but develop in the direction of full inclusion" (Olson, 2007, p. 333). The second paradox emerges when members of the community are "applying a conception of legitimacy that is premised on making predictions about the future of a political system that by definition frustrates such predictions" (Olson, 2007, p. 334). This results from the fact that it is based on the indeterminacy of dynamically forming constitutional systems.

How do these briefly described paradoxes translate into the doctrinal assumptions of liberal constitutionalism, which has dominated Western

Europe since the end of World War II and subsequent waves of democratic transitions? Primarily, they indicate the susceptibility of the democratic constitutional system to political influences, which affect its social perception. If lawmakers (who in this case should be equated with political power) do not implement Olson's concept of the minimum level of legitimacy, they will create a situation in which they become alien entities to their own citizens. These paradoxes are also observable (and perhaps especially so) in the transitional communities of Central Europe. It was there that elitist visions—though well-intentioned—were transferred to societies during the democratic transition (Skąpska, 2011) Meanwhile, on the doctrinal level, full legitimacy is presumed. In this case, free individuals establish a community that realises Rawls's vision of mutual benefits, and the entire agreement is encapsulated in a constitutional act that possesses a logical structure. It begins with a preamble containing the moral-mythical sources of its establishment (Přibán, 2007), followed by specific constitutional solutions based on the most important bond, which is the rule of law (Krygier, 2012). The moral-mythical elements inscribed in the ceremonial language of preambles form the starting point for attempting to describe the axiology of communities. It is beyond doubt that every political system can possess "some" internal axiology, but in the case of constitutional liberal democracy, it plays a particularly significant role. In contemporary Western societies, this axiology consists of the rule of law, which aims to protect the fundamental rights of individuals, including dignity, security, and other rights widely regarded as fundamental. All of this is to be protected by the rule of law, which primarily serves as a tool to limit excessive (i.e., beyond the guarantees) political initiatives. In the case of so-called established democracies, this can be supplemented by fulfilling all of the actions that mitigate the aforementioned foundational paradoxes. However, the phenomenon of disinformation, due to its novelty and highly dynamic nature, requires immediate action. This, in turn, translates into the hyperactivity of political power centres. The problem is that the only tool they know to be 100% effective is the law. Consequently, the misuse of the law carries the risk of destabilising the construction described above.

This situation forces us to ask ourselves about the social effectiveness and, consequently, the internalisation by the recipients of the legal communication formulated top-down by political elites. This is related to the fact that a lack of internalisation—rejection by members of the community—may be closely tied to a sense of social powerlessness or even the perception of elite diktat.

My main hypothesis is based on the assertion that legal counteraction to disinformation is possible only when specific systemic solutions are in place. These solutions must consider all of the elements of proper law-making, effective mechanisms for its application (with a particular emphasis on the interpretation of created or existing general and abstract norms), and the expectations of the recipients of legal communication (society/members of the political community). Otherwise, there may be a violation of the foundations of the system's existence, leading to its social delegitimisation. It is important to remember that under the conditions of functioning within a constitutional liberal democracy, this is a far more dangerous phenomenon than in non-democratic systems, which always contain an element of top-down imposition on society.

Therefore, in the first part, I will discuss the specificity of the continental legal system, with a particular focus on the already discussed elements, i.e., the possibilities and paradoxes that arise when it operates within a liberal-democratic system. The second part will address the phenomenon of disinformation as a political project that impacts legal systems established under the conditions described in this introduction. The third part will be devoted to discussing the legislative initiatives taken so far, and attempt to place them in the context outlined above.

9.2 Law as a System: The Specificity of the Continental Reality

9.2.1 The Civil Law Tradition

As George Mousourakis pointed out, "legal scholars use the term 'civil law systems' to describe the legal systems of all those nations predominantly within the historical tradition derived from Roman law as transmitted to Continental Europe through the Corpus Iuris Civilis of Emperor Justinian" (2015, p. 296). This process gained particular significance with the post-Enlightenment movement for the codification of private law, especially within the French tradition (through the Napoleonic Civil Code and its emulations) and the German tradition (through the German Civil Code—BGB—and its subsequent emulations). According to Mousourakis, the division of legal traditions (he used the term "legal families," p. 301) is based on five fundamental criteria formulated by Zweigert and Kötz: history, mode of legal thinking, legal institutions, sources of law, and ideology. The first criterion, in his view, is beyond dispute, as the entire civil law tradition derives from one core—the Roman one. The second criterion is defined by the existing "tendency to use abstract terms and, more generally, to employ a conceptual approach to legal problems" (Mousourakis, 2015, p. 301). The generality of norms allows for their application to numerous specific cases and events. The "toolbox" justifying the application of norm "A" to case "X" consists of principles for interpreting legal texts and determining the meanings of terms used in legal provisions (legal reasoning). The primary source of law in this context is the text of enacted law (in legal education, the textbook is usually based on codal systematics). Consequently, there is a clear distinction between substantive and procedural law. The extraction of historical influences has also led to a situation where even public law within the civil law tradition, though not regulating private law relations, has been heavily influenced by these normative constructions. Regarding the last element, Mousourakis stated that "ideology is the least useful criterion when distinguishing between civil law and common law, the other major legal family within the Western legal tradition. The essence of the philosophical, political, economic, and cultural foundations of law in both legal families is too similar for it to be otherwise" (Mousourakis, 2015, pp. 307–308). He also observed that this tradition continues to evolve and has influenced the shape of supranational law—namely, the European Union's law: "it has influenced the law of the European Community in structure, style of reasoning, and ethos, and continues to play an important part in the process of harmonisation of law in Europe" (Mousourakis, 2015, p. 308).

Similarly, the civil law tradition has been described by John H. Merryman and Rogelio Perez-Perdomo (Merryman & Perez-Perdomo, 2019). They have also highlighted the fact that it originates from Roman tradition and is largely based on Justinian's codification. Subsequently, through the activities of the Bolognese scholars of Roman law (glossators and commentators), its structures gradually penetrated the minds of modern lawyers. As a result, with the strong demand for codification during the Enlightenment and the nineteenth century, there were effectively no competing ideas in continental Europe. The existing law could not, in any way, compete with the high intellectual quality offered by Justinian's codification, particularly as its structures fully realized the postulate of the generality and abstraction of the norms contained in the enacted law provisions.

9.2.2 A Specific Variation of the Civil Law Tradition: the Post-Communist One

The described legal tradition, which serves as the foundation for the dominant understanding of what law generally is in Central Europe, is crucial for reconstructing the way of thinking about its role in political actions aimed at developing tools to counter and combat disinformation. However, this does not mean that the local context of Central Europe does not influence the specific details that constitute the local legal mindset. On the contrary, as noted by Alan Uzelac, a Croatian legal scholar, about fifteen years ago, he engaged in a polemic with Merryman regarding his abandonment of the term "socialist legal tradition." While Merryman equated its end with the end of the Cold War and the fall of the USSR, which institutionalised an ideological approach to law—this being the essence of the socialist legal tradition—Uzelac argued that some of its elements have survived. However, unlike Merryman, Uzelac did not focus on the content of the law and its application methods but rather on specific elements that are typical of post-communist Europe and influence the way the law functions in the former states under real socialism. He identified them as: (a) the systemic function of legal processes ("Legal Processes as the Tool for the Protection of the Interests of Political Elites" Uzelac, 2010, p. 382); (b) the characteristics of decision-making processes (he described this as "Fear of Decision-Making," Uzelac, 2010, p. 383); and (c) the impact of the material situation of lawyers (especially judges) on their attitudes.

In the first case, according to Uzelac, a characteristic feature of the post-communist legal tradition is the lawyers' orientation toward protecting the positions of elites at the expense of the demands formulated by lower social strata (fellow citizens within the political community). He attributed this to the habits of the judiciary inherited from the fallen system. He emphasised that "the connection between law and politics has existed in every legal tradition. Only in one, however, was it self-understood that the law, lawyers, and all legal elites only existed to protect the ruling elites and their political ideologies" (Uzelac, 2010, p. 382). While in the case of some countries where the ideology of real socialism was constitutionally enshrined, such a statement might be unfair to some individuals—especially some Polish advocates—who defended citizens' rights, it seems that Uzelac captured a general trend resulting from Soviet-style centralism. As a result, lawyers were primarily expected to be well-trained technicians in their discipline, capable of justifying whatever position was expected of them.

The second element determining the post-communist legal tradition stems directly from the previous one. If lawyers were trained to defend specific interests, disregarding the entire layer of universal axiology, they—by implication—were unable to step outside the rigid framework of the legal text and the technical tools of its interpretation. As a result, "most of the socialist judiciary has developed over time numerous methods aimed at evading responsibility for decision-making. Unlike the heroic figure of the common law judge who strives to contribute to legal history through prudent, brave, and well-reasoned judgments, socialist judges, in the fear of eventual retribution, always desired to remain as anonymous as possible" (Uzelac, 2010, p. 383). The civil law culture particularly favoured them in this regard. In practice, this was reflected in the reinforcement of elements constituting civil law culture, such as formalism, the associated prolongation of proceedings, extensive evidentiary processes, and excessive use of the right to appeal. As a result of such actions, even if a decision was ultimately made, the process took so long that public opinion was no longer interested in it. This allowed lawyers to remain in the shadows.

The third characteristic of the post-communist legal culture, according to Uzelac, is directly derived from the previous two. This time, it concerns the material situation of lawyers, especially judges. Given that they were merely cogs in the centralised state machinery of power, "the judiciary obviously did not enjoy a very high social esteem as the importance of its work was marginal. Monitoring the course of judicial processes, an outside observer could hardly notice the difference between the judiciary and any other clerical position in the state administration. Even the status of judges was practically the same" (Uzelac, 2010, p. 386). Nevertheless, the work of lawyers, due to its invisibility to the public (which, moreover, did not exist in the states under real socialism) and the technical nature of most cases, was highly sought after. Technicity, in this case, should be understood as the use of complex language generated internally within the profession (known in Polish legal studies as "legal language") to explain trivial disputes. As Uzelac noted, years of practising this scheme led to its consolidation. In other words, subsequent generations of lawyers trained in such a reality considered it obvious and the only acceptable one. With this view, they entered the world after the fall of communism. Therefore, he characterised it with the phrase: "old and new together: new blend old tradition" (2010, p. 388).

9.2.3 The Civil Law Tradition and Its Natural Limitations: An Overview

The main objective of this chapter is to identify the conditions under which law can be an effective tool—not just perceived as such based on wishful thinking by decision-makers—for combating disinformation. So far, we have discussed the constitutional conditions shaping Central European political communities after the fall of communism, as well as the legal tradition dominating these communities, with a focus on the practical aberrations arising from historical experience. Consequently, it is necessary to highlight the basic doctrinal and practical weaknesses that may have a real impact on the social effectiveness of legal tools used to combat disinformation, as well as to suggest potential paths for improving this situation.

The civil law tradition prevailing in Central Europe, as previously outlined, relies on a formal approach to sources of law, where the foundation of legal decisions lies in the application of the general and abstract norms contained within them, with proper interpretation and application to specific cases, in accordance with accepted principles of legal reasoning. In such processes, the legal authority must primarily assess (by determining the meanings of the words constructing the legal norm) whether a given provision applies to the case at hand and justify its decision by, among other things, comparing evidence and factual circumstances with the established "correct" meaning of the norm contained in the legal provision. The practice of applying law within the civil law tradition, due to its formalistic nature, should, in principle, be free from external influences. However, as Uzelac's reflections demonstrate, it is also shaped by the local context. What does this mean?

Formalism, as an important element shaping the discussed legal tradition, should be accepted without significant reservations in light of the analyses so far. This is evident throughout the entire stream associated with it, legal positivism. A significant influence on this comes from Hans Kelsen's theory of the basic norm. In short, it is based on the assumption that it is "the guarantor of the validity of law, not only because of its abstract construction but also because of the way it is established, closely tied to the dynamic nature of law. In other words, the Grundnorm contributes to the legitimisation of the validity of other norms in the legal order, not because their content can be logically derived from the basic norm, but because they were created [...] precisely by the basic norm"

(Huk, 2014, p. 204). Its very construction "expresses one of the fundamental intuitions allowing us, among other things, to define the concept of a normative system, its structure, and to determine the criteria for the validity of norms belonging to that system" (Stelmach, 2001, p. 63). In other words, it is the foundation of the entire legal system, and upon it, the other legal norms contained in enacted law provisions are based.

Without delving deeper into an analysis—or even an assessment of the correctness—of Kelsen's theory of norms, it is worth considering and highlighting the influence it has had on the processes of law-making and application in post-communist Central Europe. While for years, it was assumed that Kelsen entered the legal mainstream (in the Anglo-American legal tradition) due to, as Trevor Wedman put it, "his strong rejection of natural law and his advocacy for liberal democratic principles, even if such recognition has often remained somewhat reserved due to the more exotic aspects of his Pure Theory, most significantly the basic norm and Kelsen's insistence on the legal system as a system entailing validity" (Wedman, 2018, p. 112), it is precisely the mentioned basic norm that played a significant role in Central European jurisprudence. However, from the perspective of this chapter, another view of Kelsen and his theory, presented by Lars Vinx (2007), is important. According to Wedman, he rediscovered this concept, but "provided an interpretation of Kelsen according to which the Pure Theory of Law is said not to provide a normative basis or validity for the law as Kelsen explicitly claimed, but rather to explain how the law can be understood to be legitimate by ensuring the hierarchy of norms which ends in democratic discourse and mutual recognition of majoritarian rights" (Wedman, 2018, p. 112).

As I have already mentioned, the apparent logicality of the discussed theory, as well as (perhaps especially) its closed nature, seem to suggest that Uzelac's observations should not be justified in any way. However, it is worth noting the absence of a significant element that has already appeared in the introduction to this chapter: the absence of consideration of the phenomenon of social order.

The failure to consider this element may lead to the detachment of law as a system created for and by society from its source, and its transformation into an imposed system. Therefore, in contemporary democratic systems, the social legitimacy of both the constitutional order and the resulting hierarchical legal system should play a key role. Vinx, who has already been mentioned, recognised this when he titled one of the chapters of his work "Kelsen's Theory of Democracy—Reconciliation with

Social Order" (Vinx, 2007, p. 101). He saw social order as connected to legal order: "if there is to be social order, there has to be a positive legal system that successfully regulates all human conduct through the coercive rule that it creates, applies, and enforces by itself. Social order as legal order is opposed to the anarchy of a state of nature in which people are subject to nothing but their own will" (Vinx, 2007, p. 104).

The issue of social order is also a key element, for instance, in the century-old concept of the sociology of law formulated by Eugen Ehrlich, a native of what was then Austrian and is now Ukrainian Bukovina. In the English-language compendium of his concepts related to the sociology of law as a discipline (Ehrlich & Issacs, 1922), he identified the source of social order, which "rests on the fundamental social institutions: marriage, family, possession, contract, succession" (Ehrlich & Issacs, 1922, p. 131). Moreover, when considering whether it is possible to have a legal system without legal provisions (understood as enacted law), he stated that "if [...] society is older than Legal Provisions [...] it must have had some kind of ordering before Legal Provisions came into existence" (Ehrlich & Issacs, 1922, p. 132). Furthermore, he noted that social order is not written in stone. On the contrary, "it is in constant flux. Old institutions disappear, new ones come into existence, and those that remain change their content constantly" (Ehrlich & Issacs, 1922, p. 139). Nevertheless, he had already observed something that today we attempt to address through (as mentioned in the introduction) the juridification of more and more areas of social life. Specifically, he noted that "the lawgiver can, by means of his statutes, render decisions only in those types of legal cases which came to his attention" (Ehrlich & Issacs, 1922, p. 140). As a result, "it is clear why Legal Provisions cannot possibly cover the entire law" (Ehrlich & Issacs, 1922, p. 141). The areas not covered by enacted law, however, must still be regulated somehow. Their stabilisation, therefore, depended on the content of social order. This order must be considered appropriate by society and collectively adhered to. Otherwise, it would not be able to function without the petrification of its norms in the content of enacted law. Its evolution is nothing more than the adaptation of norms of conduct to a changing environment. In other words, social order is under ongoing tension to be socially recognised as a "just" normative system. Otherwise, it would cease to be adhered to. According to Ehrlich, the law not only had to be enacted, but it also had to be enforceable. For this to happen, there had to be a social will to adhere to it within social order. Therefore, the notion of justice and the social recognition of this normative system as just were key in the concept of the sociologist from Chernivtsi.

In this case, the correlation between the construction of enforceability and the aspect of social, presumably bottom-up, constitutionalization and legitimacy in contemporary constitutional liberal democracy is more than evident.

9.3 Disinformation as a Political Tool: The Experience of Central Europe

Is disinformation a tool for conducting politics? The answer should be a resounding yes. Can it be analysed solely at the level of a specific political community? Can it be analysed as a tool of international politics only? In these cases, the answers cannot be as unequivocal. Nevertheless, as I mentioned in the introduction to this chapter, my primary interest lies in disinformation as a phenomenon that originates from outside a given political community. In this context, I refer to both national and supranational structures (such as the European Union). So, what is the specific type of disinformation that individual member states and the entire Union are particularly determined to counteract? It is primarily the response to the narrative offensive that Russia (and its vassal states, especially Belarus) has been waging against the collective West since the outbreak of the war with Ukraine in 2014. This disinformation campaign has intensified significantly since the beginning of the full-scale invasion in February 2022. In this case, disinformation is not merely one of the tools for achieving political goals; it has risen to the level of a weapon within the broader framework of hybrid warfare.

Although this concept, as Mirosław Banasik (2016) rightly noted, has been known for centuries, it gained particular significance with the development of new forms of communication—the internet and, later, social media. Understood as a combination of regular and irregular actions, it defies any classification frameworks on the one hand and any legal regulations on the other. Therefore, "in countering hybrid threats, one must consider the aggressor's conviction in the consequences of its actions and the price it will have to pay for them" (Banasik, 2016, p. 41). Such consequences could be, for example, sanctions. However, when the targets are political communities based on such fundamental legal values as the presumption of innocence or the burden of proof being on the accuser, any actions by decision-makers must be based, if not on hard evidence, then at least on assertions that will be satisfactory to their internal public opinion. It is also important to note that the second solution cannot be abused in any way. This creates favourable conditions for the development of disinformation activities, which, before being falsified, begin to evolve (or, more colloquially, take on a life of their own). In the case of the actions inspired by Moscow, we can already speak of a tradition over a century old. As Aristides Mahairas and Mikhail Dvilyanski have pointed out, "its origins […] can be traced back as early as the 1920s, when Russia began using the word in connection with a special disinformation office whose purpose was to disseminate false information with the intention of deceiving public opinion'" (2018, p. 21). Through analyses of specific case studies, starting from the 1960s and ending with the ongoing campaign to weaken the Western world, they concluded that "the objective is to create confusion surrounding the true motivation behind the content and hide the origins and sponsorship by the foreign government" (2018, p. 25). Such a construction of disinformation activities naturally raises the problem of the clash of worlds that I mentioned earlier. One world is founded on precise principles, adherence to which is its vital strength, while the other derives its strength from the absence of rules.

This is evident in the context of analyses related to countering disinformation within political systems based on constitutional liberal democracy. It is particularly noticeable when the internet, especially new media, is harnessed for the purpose of achieving political goals.

As Anya Schiffrin observed, "propaganda, lies, and 'truthiness' have been around for hundreds of years and have been used by many political candidates, corporations, and religions to persuade and mislead" (2017, p. 120). However, she also pointed out a fundamental issue that, in my opinion, is crucial for understanding the scale of the challenge. Namely, she stated that "what is different today is the speed and volume of disinformation" (Schiffrin, 2017, p. 120). To confirm this, she cited the example of the 2016 U.S. presidential elections and the impact of disinformation activities on voter preferences in swing states. The profiles of organisations that were spreading disinformation favourable to the then-perceived Russian interests in 2016 also correspond to the tactical disinformation described by Mahairas and Dvilyanski. Specifically, they were: "a) Bodies thought to be close to the Russian government that circulate propaganda and disinformation with the intent of sowing confusion and distrust; b) Right-wing U.S. groups such as Breitbart; c) Groups that make money from circulating disinformation such as the notorious Macedonians profiled by Buzzfeed in the fall of 2016; d) Formal campaigns using behavioural marketing tools such as Cambridge Analytica; e) Peer-to-peer distribution networks including the far-right activists of 4Chan" (Schiffrin, 2017, p. 120). She also pointed out that political actions against disinformation will always be subject to accusations of limiting fundamental rights. For "in many cases, countries with laws against hate speech and incitement will need to find democratic ways to enforce them online so that the fight against disinformation does not become an excuse for corporate and government censorship" (Schiffrin, 2017, p. 122). The politics of combating disinformation carry real implications for democracy. Schiffrin identified the risks of instrumentalising voters by those in power, which can be exacerbated by low-quality education; the resulting problem of not equipping society (members of the political community) with tools for fact-checking; and society's approach to education, which may not show a willingness to develop critical tools (2017, p. 123).

At this point, it is necessary to ask whether disinformation politics is entirely devoid of intermediaries. In other words, is it merely a simple communication between the message sender, strongly motivated to achieve a goal, and its direct recipient? While if we limit ourselves to the narrow understanding of new media (especially social media), where the roles of creator and recipient are fluid, one might attempt to defend such a position. However, a significant portion of the information space is still filled by traditional media. In this case, they act as intermediaries and filters. What are the consequences of a lack of fact-checking at this level? In the case of Russian actions against Ukraine, it became glaringly apparent that the farther from the centre of events, the more susceptible the media were to narrative suggestions stemming from disinformation activities. This was noted by John Besemeres in his book "A Difficult Neighbourhood: Essays on Russia and Central-East Europe since World War II." Describing the state of awareness in Australian media following Russia's annexation of Crimea, he stated that: "when conflict between Russia and Ukraine first entered the Western public awareness earlier this year and Australian media were looking to bone up quickly, I noticed that a lot of the questions directed to me reflected very serious even crippling misunderstandings. I was frequently asked not to discuss the overall situation or some important development, but rather the threat posed by the neo-Nazis known to be dominant in Kyiv. Or could I please comment on and explain the reasons why Russians were in fear of their lives in Eastern Ukraine, where most people were Russian or pro-Russian and were in despair because the use of the Russian language had been banned" (2016, p. 357). In the latter part of his argument, he cited more examples of facts raised by Western journalists in relation to Ukraine, which, when deprived of proper context, became their own contradictions and, in effect, confirmations of Russian narratives. One example he cited was the issue of corruption. While its prevalence in Ukraine in 2014 cannot be doubted, resistance to the state's disintegration was one of the motivations for people protesting on Kyiv's Maidan during the Revolution of Dignity. Nevertheless, as Besemeres noted, this crucial element for understanding the context was being overlooked by Australian media. So, can traditional media become a tool of disinformation politics just as much as individuals?

Katarzyna Bąkowicz, in her analysis of the impact of disinformation on the media environment (Bąkowicz, 2024), also highlighted the particular role of social media in accelerating this phenomenon. She noted that "improving the quality of the information ecosystem is in the broadly understood public interest, and a sense of responsibility for information should be shared by all members of society. However, the greatest burden of creating and distributing reliable content rests on the media" (2024, p. 113). Moreover, she pointed out that "journalists are not immune to the temptation to create disinformative content and employ various methods to manipulate the portrayal of reality in this way" (2024, p. 113). Among the most common practices, she identified the manipulation of quotes and altering the context of information. Additionally, she rightly observed that the competition among journalists to deliver information quickly can lead to the creation and dissemination of disinformation. In contrast, the desired approach from journalists would be to act professionally, with an emphasis on independence and—as mentioned earlier in the context of individuals—fact-checking. This is especially important given that (as evidenced by the quantitative research she presented) Polish journalists demonstrate a relatively high awareness of what disinformation is and the threats it poses.

9.4 Legal Initiatives

The descriptions provided thus far clearly indicate that the application of legal instruments, particularly those rooted in the civil law tradition, necessitates juridification and a focus on a broad spectrum of issues and challenges. Areas such as new media (with a particular emphasis on their moderation), traditional media and journalistic accountability, or perhaps the precise delineation of when an individual is a creator of a message, as opposed to a recipient or merely an intermediary in transmitting false information, may require detailed legal regulation. This would necessitate, for instance, the creation of new legal definitions. However, a cursory analysis of the activities of the Polish legislature suggests that the pursuit of comprehensive legal regulation continues to elude lawmakers.

9.4.1 National Level: A Case Study of Poland

In the case of Poland, a notable delay was observed in the legislative actions taken in response to the emerging necessity to combat disinformation following Russia's full-scale invasion of Ukraine. While Poland became a frontline state (and the primary logistical hub) as early as February 2022, it was not until the middle of the following year that amendments to the law, which can be indirectly linked to efforts to counteract disinformation, were enacted. This delay in legislative response highlights a certain lag in addressing the need for legal measures to combat emerging threats in a timely manner.

At first glance, the most significant legislative change was the Act of August 17, 2023 (Journal of Laws 2023.1834), amending the Polish Criminal Code and some other acts. First of all, it increased the criminal punishment for spying, introduced a register of "spies," and introduced other restrictions related to the protection of strategic locations (including a prohibition on taking photographs). The amendment is not explicitly dedicated to disinformation, but through its subject scope, it places pressure on intensifying the penalisation of activities that involve, among other things, disinformation operations.

Since its entry into force (23.09.23 and part of the provisions on 1.10.23), the maximum sanction for espionage activities is life imprisonment (Article 130 §2 of the Criminal Code). Such a punishment can be imposed from now on a person who has provided foreign intelligence information that may cause damage to the Republic of Poland. On the grounds of other laws, such as with pension supplies, the legislator has introduced a statutory withdrawal of the right to pension for a soldier who has been convicted of the crime mentioned above (Article 10 of the Law on Pension Supplies for Professional Soldiers and Their Families, Journal of Laws 2022.2528, Journal of Laws 2023.347, Journal of Laws 2023.658, Journal of Laws 2023.1429). An analogous provision has been made for uniformed services subordinate to the Ministry of the Interior. As for the "spy registry," the Polish legislator decided to introduce provisions according to which intelligence activities by the services of countries friendly to Poland will be allowed to be carried out on Polish territory after obtaining permission from the Internal Security Agency (Article 5 of the amending law).

In addition, as part of the same amendment, the Polish legislature introduced changes to the Law on Defense of the Homeland (Journal of Laws 2022.2305, Journal of Laws 2023.347, Journal of Laws 2023.641, Journal of Laws 2023.1615). From the fall of 2023, under penalty of arrest or a fine, it is prohibited to photograph objects that serve a sensitive function for state security and are labelled by a sign saying "no photography." However, the legislator did not specify which objects are at stake. According to the provision's text after the amendment, it refers only to three categories of objects: those that are particularly important for the security or state defence, objects of the Ministry of National Defense that do not belong to the first category, and objects of critical infrastructure.

In summary, the Polish legislature in 2023 did not carry out comprehensive legislative measures directly dealing with the disinformation issue. Instead, it carried out an initiative to amend the current legal framework. New formal restrictions were introduced, and the extent of the criminal sanction was increased. It is important to emphasise once again that the provisions that have been amended can be applied to disinformation activities in the process of law enforcement. As a result, it should be concluded that in 2023, the Polish legislator focused on filling in the normative and interpretative gaps in the legislation that had been binding until then.

9.4.2 The Supranational Level

At the supranational level, the most detailed regulation currently is the Digital Services Act (DSA), formally known as Regulation 2022/2065 of the European Parliament and the Council of October 19, 2022, on a Single Market for Digital Services and amending Directive 2000/31/EC (DSA) (L 277/1). The fact that this is a Regulation, rather than a Directive, underscores the significance of the subject matter. Unlike Directives, Regulations are directly effective in all Member States and do not require transposition. In other words, they are uniformly applicable across the entire European Union as adopted through the ordinary legislative procedure. A comprehensive analysis of the detailed provisions would require a separate study; however, it is worthwhile examining the rationale for its adoption as articulated by the EU legislator.

Among the 189 recitals justifying the need for this Regulation, the term "disinformation" appears thirteen times: in recitals 2, 9, 69, 83, 88, 95, twice in recital 84, and in recitals 104 and 106. Furthermore, the first sentence of the first recital states that "Information society services, and especially intermediary services, have become an important part of the Union's economy and the daily life of Union citizens." This suggests that EU legislators are aware of the increasing importance (and even vital nature) of the electronic circulation of information in the modern world.

In the second recital, one of the reasons for adopting the Regulation is identified as the independent regulatory attempts by Member States: "Member States are increasingly introducing, or are considering introducing, national laws on the matters covered by this Regulation, imposing, in particular, diligence requirements for providers of intermediary services as regards the way they should tackle illegal content, online disinformation or other societal risks," which, according to the EU, could negatively impact the internal market regulated by the Treaty on the Functioning of the European Union (TFEU). Therefore, recital 9 states that "This Regulation fully harmonizes the rules applicable to intermediary services in the internal market with the objective of ensuring a safe, predictable, and trusted online environment, addressing the dissemination of illegal content online and the societal risks that the dissemination of disinformation or other content may generate." In subsequent recitals addressing the phenomenon of disinformation, the problem of manipulating the recipient, for instance, through content targeting, is highlighted (recital 69). Consequently, "manipulative techniques can negatively impact entire groups and amplify societal harms, for example by contributing to disinformation campaigns or by discriminating against certain groups. Online platforms are particularly sensitive environments for such practices and they present a higher societal risk." Recital 83 explicitly states that organized disinformation campaigns can negatively impact the well-being, including mental health, of EU citizens and residents. They may also cause unrest that harms particularly legally protected groups, such as minorities.

Recital 84 directly references the responsibility of professional entities, specifically focusing on internet service providers: "When assessing such systemic risks, providers of very large online platforms and of very large online search engines should focus on the systems or other elements that may contribute to the risks, including all the algorithmic systems that may be relevant [...] Such providers should therefore pay particular attention to how their services are used to disseminate or amplify misleading or deceptive content, including disinformation." This clearly indicates the legislator's expectation that professional entities engaged in this field will assume a portion of social responsibility. Recital 88 is formulated in a similar tone, explicitly stating that online platform providers should, among other actions, "consider awareness-raising actions, in particular where risks relate to disinformation campaigns." The following recital also emphasizes the social responsibility of entities with professional knowledge, specifically targeting advertising service providers (recital 95).

Finally, recital 104 explicitly suggests that disinformation activities undertaken online may undermine the axiological foundations of the European Union: "In particular, risk mitigation measures concerning specific types of illegal content should be explored via self- and co-regulatory agreements. Another area for consideration is the possible negative impacts of systemic risks on society and democracy, such as disinformation or manipulative and abusive activities or any adverse effects on minors. This includes coordinated operations aimed at amplifying information, including disinformation, such as the use of bots or fake accounts for the creation of intentionally inaccurate or misleading information, sometimes with the purpose of obtaining economic gain, which are particularly harmful for vulnerable recipients of the service, such as minors."

9.5 Conclusions: Lawmaking and Law Application in the Case of Disinformation—In Search of Balance

The dynamic nature of disinformation, particularly when employed by external actors targeting the European Union and its individual Member States, especially those in Central Europe, necessitates extraordinary and unconventional solutions. The paradox of this situation lies in the fact that Russia's adversary, the collective West, is grounded in principles (including legal principles) that preclude decision-makers from taking shortcuts. At both the level of establishing a political community (especially in relation to the specific values typical of constitutional liberal democracies) and in the expected practice of lawmaking and law enforcement, prioritizing ad hoc solutions, even if necessary, over systemic actions based on clearly defined rules can lead to the disintegration of the entire democratic system. This, in turn, may result in its social delegitimization.

The situation in Central Europe is even more complex. Within the civil law tradition, particular importance is attached to the text as the sole source of law. Legal operations (and the legal norms they contain) must also be conducted based on accepted tools of legal reasoning. This approach to law can certainly be seen as protecting social order from excessive politicization. However, in the context of post-communist Central Europe, these principles have been elevated, within the internal practice of professional fields, to the status of fundamental principles. While one can speak of generational change and the evolution of attitudes resulting from the integration of various Central European political communities into supranational structures (such as the EU), effective legal solutions (as well as concrete actions) to combat systemic external disinformation are needed here and now.

In what situations would the application of law to combat disinformation be socially effective without exposing the ruling authorities to accusations of hypocrisy? Ideally, they would need to consider all the factors necessary to connect the laws being created and subsequently applied with social order. In other words, they would need to accurately reconstruct what the EU's supranational community considers a just solution to the problem at a given time and incorporate this into the legal provisions. Why the supranational community and not the societies of individual Member States? The motives for adopting the Digital Services Act, as discussed in the previous paragraph, clearly indicate that the European Union is concerned about the potential fragmentation of the common market, which is fundamental to its functioning. Consequently, this task would be much more challenging.

What solutions could be implemented to achieve the goal while avoiding the trap of wishful thinking, which may hinder attempts to find immediate solutions? Primarily, it is essential to undertake soft measures, including awareness-raising and educational initiatives. Once again, reference must be made to EU legislators, who have explicitly emphasized the aspect of social responsibility among professional entities. Of particular importance is raising awareness of threats and disseminating professional knowledge about disinformation to other entities involved in combating it, such as those involved in law enforcement. In this way, without the need for deep, multi-generational internal reforms (which may be especially crucial in the frontline post-communist Central Europe), these entities would be able to apply the existing laws to new cases without undermining the foundations of their existence in a constitutional liberal democracy.

References

Bąkowicz, K. (2024). Świadomość i wiedza na temat dezinformacji wśród polskich dziennikarzy newsowych. In K. Stasiuk-Krajewska & M. Weznel (Eds.), *Dezinformacja w czasach kryzysu* (pp. 107–140). Wydawnictwo Adam Marszałek.

Banasik, M. (2016). A changing security paradigm. New roles for new actors—The Russian approach. *Connections, 15*(4), 31–43.

Besemeres, J. (2016). *A difficult neighbourhood: Essays on Russia and East-Central Europe since World War II.* ANU Press.

Ehrlich, E., & Issacs, N. (1922). The sociology of law. *Harvard Law Review, 36*(2), 130–145.

Hardin, R. (2001). Law and Social Order. *Noûs, 35*(s1). https://doi.org/10.1111/0029-4624.35.s1.3.

Huk, M. Z. (2014). Norma podstawowa w Czystej Teorii Prawa Hansa Kelsena. *Ethics in Progress, 5*(2), 204–214.

Krygier, M. (2012). Rule of law. In M. Rosenfeld & A. Sajó (Eds.), *The Oxford handbook of comparative constitutional law* (pp. 233–249). Oxford University Press.

Mahairas, A., & Dvilyanski, M. (2018). Disinformation—Дезинформация (Dezinformatsiya). *The Cyber Defense Review, 3*(3), 21–28.

Merryman, J. H., & Perez-Perdomo, R. (2019). *The civil law tradition. An introduction to the legal systems of Europe and Latin America* (4th ed.). Stanford University Press.

Mousourakis, G. (2015). *Roman law and the origins of the civil law tradition.* Springer.

Olson, K. (2007). Paradoxes of constitutional democracy. *American Journal of Political Science, 51*(2), 330–343.

Přibán, J. (2007). *Legal symbolism. On law, time and European identity.* Ashgate.

Rawls, J. (1971). *A theory of justice.* Harvard University Press.

Schiffrin, A. (2017). Disinformation and democracy: The Internet transformed protests but did not improve democracy. *Journal of International Affairs, 71*(1), 117–126.

Skąpska, G. (2011). *From 'Civil Society' to 'Europe'. A sociological study on Constitutionalism and Communism.* Brill.

Stelmach, J. (2001). Norma podstawowa. In J. Stelmach (Ed.), *Studia z filozofii prawa, t. 1.* Wydawnictwo Uniwersytetu Jagiellońskiego.

Uzelac, A. (2010). Survival of the third legal tradition? *Supreme Court Law Review, 49*, 377–396.

Vinx, L. (2007). *Hans Kelsen's pure theory of law: Legality and legitimacy.* Oxford University Press.

Wedman, T. (2018). The rule of law. Toward a positive conception of state. In P. Stekeler-Weithofer & B. Zabel (Eds.), *Philosophie de Republik* (pp. 111–127). Mohr Siebeck GmbH and Co..

The manufacturer's authorised representative in the EU is Springer Nature Customer Service Centre GmbH, Europaplatz 3, 69115 Heidelberg, Germany. If you have any concerns regarding our products, please contact ProductSafety@springernature.com

Printed and bound by CPI Group (UK) Ltd, Croydon, CR0 4YY

01/10/2026

02225913-0003